BROOKLYN BRIDGE PARK

2

MICHAEL VAN VALKENBURGH ASSOCIATES

BROOKLYN BRIDGE PARK

WITH ESSAYS BY JULIE BARGMANN AND AMANDA HESSER

A PARK CAN NO LONGER BE SEEN AS "A THING-IN-ITSELF," BUT RATHER AS A PROCESS OF ONGOING RELATIONSHIPS EXISTING IN A PHYSICAL REGION—THE PARK BECOMES A "THING-FOR-US."

ROBERT SMITHSON
FREDERICK LAW OLMSTED AND THE DIALECTICAL LANDSCAPE, 1973

EAST FLATBUSH
MIDWOOD
BENSONHURST
BOROUGH PARK
PROSPECT PARK
SUNSET PARK
SOUTH SLOPE
PARK SLOPE
GOWANUS
BOERUM HILL
DOWNTOWN BROOKLYN
FORT GREENE
BQE
DUMBO
JAY STREET
DUMBO
MAIN STREET
MANHATTAN BRIDGE
JOHN STREET

STATEN ISLAND
NEW YORK HARBOR
GOVERNOR'S ISLAND
RED HOOK
COBBLE HILL
BROOKLYN HEIGHTS
ROOKLYN QUEENS EXPRESSWAY (BQE)
PIER 6
PIER 5
PIER 4
PIER 3
PIER 2
PIER 1
MANHATTAN
EMILY ROEBLING PLAZA
EMPIRE FULTON FERRY
MAX FAMILY GARDEN
BROOKLYN BRIDGE
EAST RIVER
BROOKLYN BRIDGE PARK

FOREWORD

JULIE BARGMANN

You can tell a park is great when someone absorbs it into their everyday life. The things that make up a park—hills, trees, paths, playgrounds, lawns, views—cease to be mere things when a person melts into them. How does that happen? At what point is a parkgoer taken in by the whole of a built landscape? How does anyone accept an invitation to play, rest, or "get lost" (in the best sense) in a park?

I believe the answer derives from the artful imagination and technical prowess of the designers and their collaborators. A great designer's deep-seated motivation is to use every *thing* to create the calculated complement to a person's urban existence. Anticipating the effects of spending time in a park requires an intrinsic ability to forecast. For decades now, Michael Van Valkenburgh and his crew have demonstrated heartfelt humanism, hell-bent on making landscapes for those hoping for an inward journey guided by site-specific cues and joyful invention.

Michael and the design team use the term "getting lost" to conjure up the sense of immersion they imagined for the park, spending hours wandering on the former industrial site. As he has always done and mentored others to do, Michael scrutinizes landscapes—every landform, every plant, in storms and in stillness—with an intense focus on how another person might experience, in this case, the enormous, constructed riverside surrounded by iconic landmarks and unfurling vistas. You can tell that Brooklyn Bridge Park has grown out of this generous stimulant, with beautiful elements elastic enough to broadly connect with a complex collection of eager park visitors, humans and nonhumans. The elasticity imbues strong evocative forms that have been unlocked. Each form, each place, defers to the parkgoers to shape them into comingled intention. Of course, many other parks strive to do the same, yet so many remain empty compositions, competent in their place-making but with no one there to make the place.

Brooklyn Bridge Park attracts parkgoers who, knowingly or not, take the temperature of the landscape as well as their own, and then settle into it. The number of distinctive places within the park is dizzying, much like the feeling induced by staring at the perfectly singular Brooklyn Bridge itself. Anticipating this, authentic experiences have resulted from understanding form and program as vehicles. Very simply put, the genius of this landscape comes from the ambition to make a genuine, forcefully benevolent people's park. This is why Brooklyn Bridge Park is a great park. A park worthy of this story of its making.

The designers' inner convictions to make a people's park must not have been easy to sustain over the twenty-five years of the landscape's creation. Not when an army of self-interested parties vied to grab their piece of the abandoned industrial floodplain. How did the designers keep their own impetus and instincts intact to ensure all would benefit, the heard and the unheard? Frederick Law Olmsted did this with his powerful words, writing letters as Central Park superintendent to promote investment in the park's democratic ideals for the health of urban dwellers. What about the park's Mall, one of its best-known features, matters more? Is it the axial form, the landscape type that Olmsted internalized while visiting Europe? Or is it the multitude of newcomers to a great park, folks strolling in the shade of heroic trees, for the first of many visits? There, the vast array of individuals pass each other, day after day, with increased tolerance and understanding for the other, at peace. A century and a half later, the formula still works, just as we need it to.

If Olmsted wrote eloquent letters, Michael Van Valkenburgh Associates (MVVA) made their design intent legible to the public through elegant site models. At a whopping fifty feet long, the Brooklyn Bridge Park maquette stretched across an entire floor of the firm's office (in Union Square back then). After such a long political haul, the commanding three dimensions sent a clear message: Brooklyn deserved this park. Adversaries and advocates, skeptics and enthusiasts looked upon the depiction of the landscape itself: now it had the last word. That huge model displayed both a clear design and a social imperative: fair play among unapologetic forms. The deliberate unfolding of a park based on nature as a platform for democracy steadied the process of hearing thousands of desires and demands from the community. The designers' listening was more like ingesting, nourishing their own motivation to freely interpret the needs of the silent, suppressed site. For example, the sounds now coming from the postindustrial "Picnic Peninsula" are not unlike those of Central Park's Mall, operating similarly in Brooklyn, where equity and compassion barbecue side by side.

An opaque monoculture, the fallow site required X-ray vision and a telescope. Careful looking led to finding cues surfacing as promises not to tame but to amplify. Curating the complicated structures and surfaces involved calibrated subtraction, addition, and leaving well enough alone. MVVA's primal and primary impulse was to thicken the relief of the level plane, diving down between the piers to the heartbeats in the muck, projecting up in spite of the screaming Brooklyn-Queens Expressway to the Brooklyn Heights plateau. The designers' nerdy love of science and technology took it from there. The industrial flats became the fulcrum for carving a laminated water's edge and for piling earth up high, very high. And then the test fits began—every form and path honed, time and time again, for immersion's sake. Those operations were the means. The end is the beginning of many stories from park enthusiasts who needn't apprehend the engineering and ingenuity in order to recollect their heightened day in the magnified landscape.

One has only to look at Van Valkenburgh's recent book, *Designing a Garden*, to sense the motivation for creating landscape experiences through montages of textures and light; duration plays its part. At Boston's Isabella Stewart Gardner Museum, Michael's lifetime of finding nuances translated into tangled paths for one to wander aimlessly, to find the freedom to get lost. *Designing a Garden* is less a documentation of how the garden was made and more a soulful story to elucidate the garden's painterly aim to transport. Similarly, this story of making a park will reveal how those instincts born in a garden grew generously in Brooklyn Bridge Park.

With a multitude of named spaces filling its eighty-five acres, Brooklyn Bridge Park could still appear wonderfully overcrowded with program. Rather than a checklist of activities, the mosaic of experiences collaged as readings of the site are magnified by sculpted water, landform, and planted form. Recognizable pieces like playgrounds, sports fields, and picnic areas are all infused by the sensory landscape associations that make them more than dutiful program. The backgrounds of the designers inform every aspect of the spaces. For instance, the "Swing Valley" playground, as recounted by MVVA partner Matt Urbanski, contains a profuse number of fluffy-headed hydrangea, the colorful garden variety etched into Matt's childhood memories of the New Jersey Shore. His aim was to hand off that flowering recollection to a city kid who may never get to the Atlantic. "Nature draws people out of themselves. It allows them to step out of themselves and be children, to be wondrous," Urbanski said.

In contrast to the intricacy of play, every massive piece of the context from the harbor to the bridge keeps beat to the metronome of activity. It's masterful, really, to speed up unclad piers with play, to pace stacked stone with a prolonged view, to set a slow timer on the tide with spiraling boulders. Rewinding to the site model, picture the designers taking a small-scale figure and placing it in each part of the park, taking its pulse. Anticipating immersion has been a long-standing design philosophy and practice for MVVA. No neutral spaces allowed, nearly silent maybe, but not neutral. Every space, every experience is charged in a distinct way. The meaning of the words recreation, rest, and play come to life in Brooklyn Bridge Park.

Evenhanded plus formidable design motivations comprise the why and how Brooklyn Bridge Park emerged and continues to evolve. The loved and lauded landscape will grow socially and ecologically, tethered together emotionally by the thousands of parkgoers who find their place within it every day with ease. Tourists will always gasp at the spectacular views and then refocus to see the people who shape the park, etching the earth, rock, water, plants, and every living thing with their life-affirming rituals. As you read this book, I urge you to look underneath the stories for the rigorous instincts and generous spirit in the making of a magnanimous landscape.

INTRODUCTION

MICHAEL
VAN VALKENBURGH

Sometimes a design firm and a project are so closely aligned that it's hard to imagine one without the other. For MVVA, that project is Brooklyn Bridge Park. It is at the heart of who we are. In 2008, our New York office moved from Union Square in Manhattan to Brooklyn Heights to be closer to the site, joining thousands of Brooklynites who would soon claim this place as their own.

This book tells the story of the planning and design of eighty-five acres of public space reclaimed from a defunct commercial waterfront on the East River. Each section recounts the challenges, insights, and innovations concerning a key element of the project, building on more than a decade of reflection. Improved access and newly designed urban connections to the waterfront bring the energy of the city into the formerly isolated site. Massive constructed topography offers, in equal measure, tranquil spaces and dramatic vistas. Abundant plantings, from grove to meadow, strive to be big-hearted and ecologically high performing. A re-engineered water's edge honors the dynamism of the tidal East River. Finally, a photo essay celebrates being in the park, shifting attention from the design to the visitors who give it vitality and meaning.

In 1999, about the time of the first public meeting concerning the future park, Matt Urbanski and I took a long walk with Elizabeth Barlow Rogers in Central Park. She had initiated a new chapter in urban park history by organizing the Central Park Conservancy in the 1980s, helping raise millions and developing a "management and restoration" plan that prioritized continuous care for the landscape as much as capital spending. As a result, today Central Park is in the healthiest condition it has been in since the early twentieth century. These accomplishments did not come easily. Betsy told us, "If you are going to design a park in New York City, you're going to have to learn how to bend." Before long, everyone on our team came to understand what she meant. Bending isn't giving in to the naysayers or compromising your ideals. Instead it is a form of empathy, the ability to adapt to the needs and desires of stakeholders at least as much as you would expect them to embrace your ideas. It also means letting the form and function of a landscape emerge organically from the site rather than imposing preconceived ideas upon it. Bending requires the humility and determination to create something richer and more authentic than anything that could spring from your imagination alone. We inherited an incredibly complex site and a project subject to many steadfast and sometimes conflicting interests, as well as glaring scrutiny. Working with the Brooklyn Bridge Park Corporation, a self-governing entity of the City of New York, we crafted a design process in collaboration with the community that was bolstered by the expertise of thirty-five specialized consultants. With Betsy's advice, we learned to be supple in our approach to the park's design and engaged its stakeholders with patience, empathy, and optimism.

These days, the concept of urban waterfront parks as flood infrastructure is familiar to landscape architects, ecologists, and engineers, many of whom are doing excellent work to protect coastal and riverine cities. Twenty-five years ago, few precedents existed for such public spaces at the scale we envisioned. Reshaping such a considerable portion of Brooklyn's shoreline allowed us to imagine a more adaptive relationship with the water. This meant replacing acres of impervious pavement with a verdant threshold that would allow floodwaters to rise and recede within the park, thriving amid the extreme weather that—given the climate crisis—is becoming a more regular part of city life. The park is a model for resilient waterfronts because it was designed not only as a piece of infrastructure but also as an environment where people are able to interact with the ever-changing natural world.

We recognized that taking a purely defensive posture toward the river and the forces of climate change would result in a park that fell short of the community's aspirations. People need to have positive connections with natural systems in their everyday lives. Building this into designed spaces is how cities can grow public investment in resilient, landscape-based solutions, versus traditional engineering. And just as urgently, it's how designers can help claw back the basic nourishment that nature provides when city-dwellers have access to it. The salt marshes, beaches, and thousands of feet of reshaped banks that constitute the new shoreline will accept future floods as they did during Hurricane Sandy in 2012. In the meantime, the transformed water's edge continues to mark the tides and seasonal habits of shorebirds for visitors just steps away from the daily life of the park.

Public in more than just the legal sense of the word, the park would be as responsive as possible to the needs and desires of the people who use it.

This book details the scientific precision underlying many of the park's constructed ecologies. But it was never our purpose to put that work itself on display. In fact, we did not aim for park visitors to be preoccupied by, say, the technical achievement of transforming cracked concrete into a thriving wetland. Better to have people walk past a line of turtles basking on a log in a pond and feel the quiet pleasure of belonging among other living things. For this kind of encounter to transcend novelty and become part of daily life, we had to ensure nothing in the park's organization felt prescriptive and that every experience could be meaningful because it felt ordinary—which is not to say lacking in power or surprise. We needed a place where your kid can drop their ice cream cone, it gets cleaned up, and no one minds much. And when a visitor is struck by some moment of singular beauty, we wanted it to be about sharing a universal human connection to landscape rather than as a connoisseur of design.

Informality is a recurring theme throughout this book. We prioritized frugal construction methods and materials, reducing costs and carbon emissions while ensuring that no element was too precious to transform in the face of evolving constraints. Vibrant, unmannered plantings that were inspired by the region and reclaiming local materials helped us maximize durability and economy. These same strategies allow all kinds of visitors to connect personally with the site, without sensing the hand of a designer.

Through planning and good fortune, MVVA was uniquely positioned to embrace an adaptive way of working. First, we were nearby. Moving our office so close to the site was an early investment in community relationships that continues to pay dividends today. We could easily meet with local stakeholders in their own neighborhoods, and as the pace of design intensified, we visited the site as often as needed, constantly creating full-scale mockups and testing ideas under real conditions. Second, our team of consultants not only had exceptional skill but shared our commitment to experiment, iterate, and trust intuition. We worked with people who loved the idea of this park as much as we did.

As the park's planner and designer for over two decades, we have relied on the support of four successive New York mayors and myriad other municipal leaders and community members. The efforts of Mayor Mike Bloomberg and Deputy Mayor Patti Harris to transfer ownership of the land from the State to New York City were pivotal in the success of the park. It would be difficult to overstate the dedication and excellence of the Brooklyn Bridge Park Corporation as a client. Like MVVA, the organization has evolved specifically to meet the demands of this project, and we continue our deeply rewarding collaboration with ongoing improvements to the park.

From the beginning, our highest ambition was for the park to be a democratic landscape. Public in more than just the legal sense of the word, it would be as responsive as possible to the needs and desires of the people who use it, not just by providing amenities but by being relaxed enough for people to bring their whole selves there. We hope what follows will shed light on how one park was designed to be democratic at its core, and that readers find themselves in that wonderful space between the first expression of an idea and a living, everyday landscape—one that is best measured in basketball games, summer cookouts, and late-night strolls with a restless newborn.

OCEAN

PIER 3
OCEAN KAYAK

BEFORE THE PARK

BEFORE THE PARK

ABOVE *A pin produced by community advocates for the park, 1980s.*

BELOW *Murals from the "This Is Your Park" campaign, painted on the site's existing pier sheds to signal the change to come, 2007.*

For our first visit to the waterfront in 1999, we gathered not far from the Brooklyn Bridge. Right away, we noticed that there were two completely different sets of conditions to the north and south. Both were once a key part of the borough's commercial waterfront, and both were in dire need of transformation. Although we could have filled all eighty-five acres with an entirely new landscape, we were committed to building on the complexity of everything we inherited—the open space, scores of empty buildings, precious views caught through monumental infrastructure—to make a park that felt like it belonged where it was. The site was not at all empty; it was overwhelmingly full.

To the south of the Brooklyn Bridge lay a mile of land with barely any sign of human activity. The most prominent structures were the five giant piers that were built out into the East River in the 1950s by the Port Authority of New York and New Jersey. They were still covered by long-vacant steel sheds, but we knew even then that the piers could be perfect for a park. They came with jaw-dropping views of Lower Manhattan and the New York Harbor, and it seemed almost impossible for a park to fail in that context. On the shore—known as the uplands—half the area was cluttered with parked cars, cinderblock storage sheds, and derelict warehouses, in addition to the roar of the Brooklyn-Queens Expressway (BQE), so loud it had a physical presence.

North of the bridge, there were a couple of nineteenth-century brick buildings in various states of disrepair, left vacant for dozens of years. There were also a few simple open spaces on the shoreline, Empire Fulton Ferry State Park and Main Street Park, that were well used by the already growing DUMBO neighborhood. As we explored the city edge that first day, we were told by a representative of New York State, which owned the land, that the historic Tobacco Warehouse was slated for demolition. We asked them to wait until we could discuss alternatives—our first act of design on the site. Today, the shell of that brick structure frames the Max Family Garden at the St. Ann's Warehouse performing arts center.

By the 1980s, the future of the site was a battle for the opposing interests of many kinds of New Yorkers. As soon as the piers were abandoned by the Port Authority, the waterfront became contested territory. One group of Brooklynites began promoting a future "Harbor Park" later

As soon as the piers were abandoned by the Port Authority, the waterfront became contested territory.

in that decade. But the Port Authority continued to pursue a high-rise commercial development. "A team of prominent Manhattan developers is poised to build housing, a hotel, shops, and a marina on the East River piers at the foot of Brooklyn Heights," the *New York Times* reported in 1989. The story noted that a "tentative agreement" had been reached and "would appear to foreclose the possibility of building a riverfront park on the piers."

It took another decade of grassroots organizing before reaching an agreement to construct Brooklyn Bridge Park on the land donated by the Port Authority. In 1999, MVVA had joined a master planning team led by the economic development consultants HR&A Advisors and Toronto-based planning firm Urban Strategies. Our master plan, released in 2000, was built around the city's bold and controversial precondition that the park cover the costs of its own maintenance and operations with revenue captured from private development on the site. This defining move ensured that the park's upkeep would be funded in perpetuity and guaranteed public financial support for the park's construction. Mayor Michael Bloomberg loved the idea of the park so much that he brokered a deal for New York City to take ownership from the State, placing the new public land in the hands of the people who had fought for it.

Following the second master plan in 2005, which focused more on design and was led by MVVA, ground was finally broken in 2008. The rest of the park would ultimately require thirteen years to build in phases. An unexpected consequence of taking so many years to build the park was that users got to experience the completed sections of the landscape alongside the baffling complexity of the pre-park docklands—giving people a taste of how we felt on that first visit. This gave rise to a marvelous tension between the site's past and future as sections of the park sprung up like oases in the postindustrial desert.

Pier 1, just south of the Brooklyn Bridge, and the Pier 6 uplands, almost a mile south, were completed in 2010, when the entire center of the park was still undergoing demolition. The president of the Brooklyn Bridge Park Corporation at the time, Regina Myer, immediately recognized the importance of cultivating the image of the park in its entirety. She hired contractors to link the completed phases with a temporary asphalt path and affixed renderings of future phases to the chain-link fences that separated the walkway from active worksites. She also installed salvaged cobra head light fixtures from one end of the path to the other. The scrappy quality of the walkway made it popular with locals, and temporary amenities like outdoor art exhibitions and a pop-up pool created destinations within the unfinished interior. By then, most of the buildings south of the bridge that were slated for demolition had come down, only to be replaced by temporary nurseries, stacks of granite, mounds of soil, and massive earth-moving equipment. Together these comprised an insider's view of the landscape-in-progress, an invitation to take in the evolution of the park they would someday call their own.

As designers, our work always begins with embracing each site's limitations. We thrive on the potential to create inviting landscapes from the most improbable conditions. To us, there is no bad site. But knowing the contentious history of the land and the wildly different ways people had imagined it over the years, we could not be sure everyone else would accept that position. Shortly before groundbreaking, the whole of New York City was inspired by the artist Olafur Eliasson, who, with support from the Public Art Fund, placed two monumental *New York City Waterfalls* in the future park. One installation, just beneath the Brooklyn Bridge on the river side, consisted of a ninety-foot-high scaffold through which river water was pumped out of the top, creating a cascade at a surreal scale. The second *Waterfall* pressed up near the BQE, drowning out the noise from the expressway and appearing to flow from the Brooklyn Heights Promenade above.

Waterfalls ignited hope in the city about the work to come on the waterfront. The unfussy structural supports paired with the ethereal grace of falling water felt like the relationship between found structures and new landscape that we were working to cultivate in the park. Over the years, as visitors navigated the gritty interim conditions that gave way to the finished park, we have consistently seen them embrace landscape experiences that revel in the unresolved, unrefined, layered history of the site. We are grateful to have been able to offer, rather than obscure, those temporary wonders.

THE SITE AS FOUND

Sprawling midcentury pier sheds occupied nearly a quarter of the site; the National Cold Storage warehouse dominated the inland side of Pier 1.

PREVIOUS PAGES *Site images, 2007.*

RIGHT *Pier 1 looking toward the Brooklyn Bridge, with the shed on the left and the National Cold Storage warehouse on the right, 2007.*

NO PARKING ANY TIME

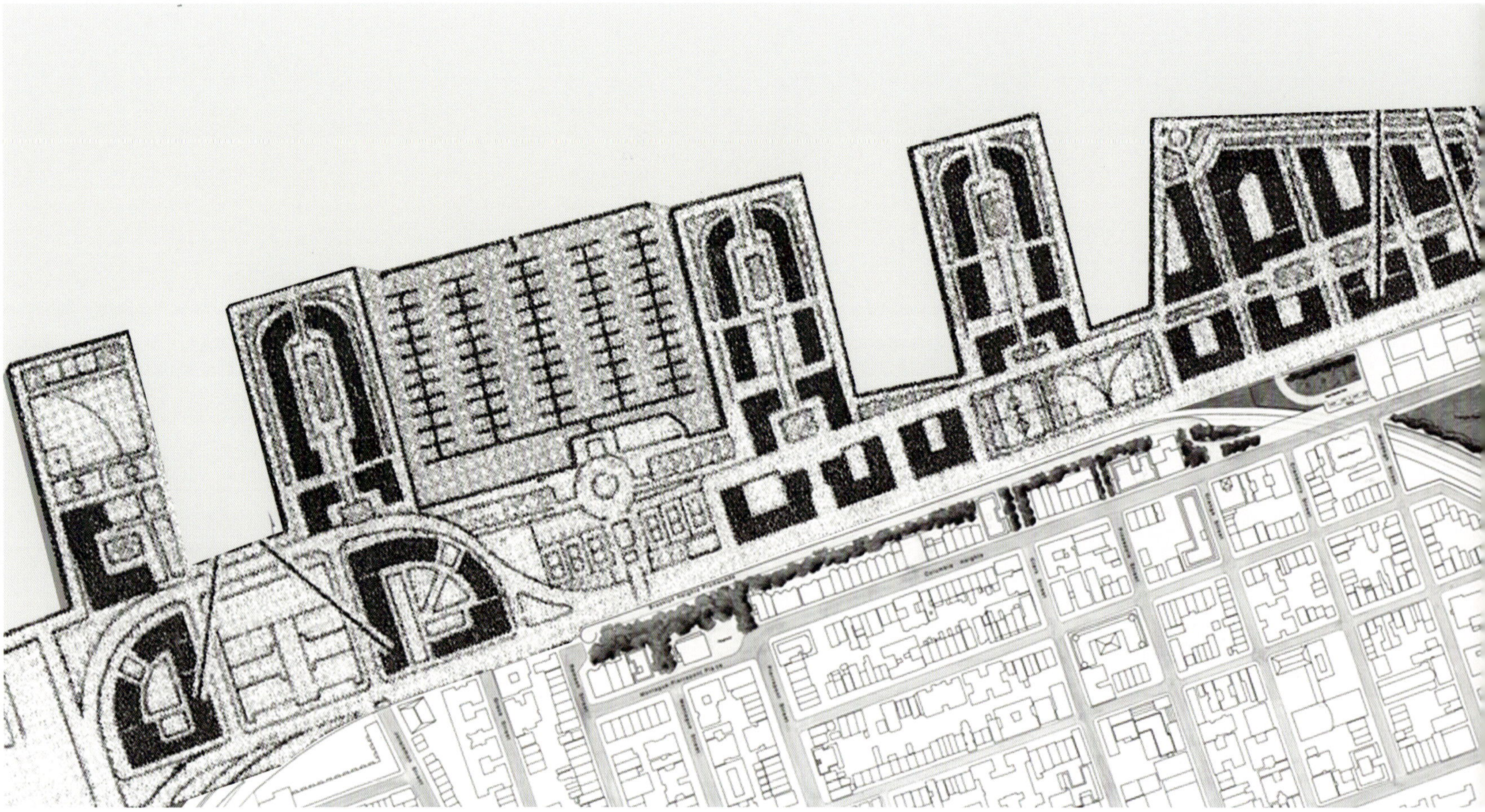

PORT AUTHORITY OF NEW YORK AND NEW JERSEY DEVELOPMENT PLAN, 1985

BROOKLYN BRIDGE PARK PLAN, MVVA, 2023

FROM A PROPOSED PRIVATE DEVELOPMENT TO A BELOVED PUBLIC PARK

The Port Authority's plans to sell the site south of the Brooklyn Bridge to developers in the 1980s galvanized local advocacy for a public park on the site.

LEFT *Plan for high-rise apartment buildings proposed by the Port Authority of New York and New Jersey, 1986.*

BELOW *Current plan for the contiguous 1.3 mile riverfront park, 2023.*

BEFORE THE PARK, 2007

ONE
CASINO ST. CHARLES

BROOKLYN BRIDGE PARK, 2021

LEFT *Existing edge condition at Pier 1, 2007.*

RIGHT *The promenade and Granite Prospect at Pier 1, 2014.*

LEFT *The cargo shed and parking lot at Pier 2, 2007.*

RIGHT *A performance on Harbor View Lawn at Pier 1, 2012.*

LEFT *The interior of the shed at Pier 2, 2007.*

RIGHT *Basketball courts at Pier 2, 2014.*

21

Pier 2 and the Spiral Pool nearing completion beside staging areas in the uplands, 2013.

DESIGNING THE PARK

DESIGNING THE PARK

In 1999, nearly a decade before construction of the park began, the transformation of Brooklyn's postindustrial waterfront into a public place was unimaginable to all but a small group of ardent supporters. The site's bleak condition and tenuous connections to the city made it impossible to imagine what the place could become. That challenge fell to MVVA. We began by consulting with economic development specialists HR&A Advisors on their Illustrative Master Plan released in 2000, and then creating a conceptual design in 2003. Soon after, MVVA led the development of the *Brooklyn Bridge Park 2005 Master Plan: A Framework for Design*, which guided implementation through the completion of the park in 2021. Our primary task was to translate the passion of park advocates into drawings, models, and descriptions that helped others see the site's potential.

Brooklyn Bridge Park was planned at a time when more and more urban design projects were being led by landscape architects. Civic leaders were learning the value of planning practices in which the local landscape drives the form of the built environment as an alternative to starting with architecture and fitting the public realm into leftover spaces. At this point, the firm had experience in bringing together the expertise of ecologists, scientists, engineers, architects, and artists, and we were comfortable navigating governmental regulations at the city, state, and federal levels. This interdisciplinary work was only half the reason MVVA was selected to direct the final planning and design effort; the other was our commitment to a design process that found inspiration in the site's existing conditions—no matter the challenges.

The failing infrastructure—eroding docks, empty warehouses on oversized piers, almost a mile of collapsing seawalls—could have been approached as a list of problems to solve. But we understood the site in a more positive light, that is, as a collage of structures and spaces pointing back to the waterfront's industrial history and forward to a range of new and largely untested landscape possibilities. Rather than clear away the so-called challenges before getting to work, MVVA used the site's inherent limitations to create a vision for a singular public landscape.

Most parks meet basic public needs by providing a combination of dedicated facilities and flexible spaces. But the landscape of a great park is an experience in itself. The sensations brought about by plants, water, weather, and topography remind us of our primordial connection to the natural world. For more than 150 years, New Yorkers have lived among terrific examples, chiefly Central Park and Prospect Park. In the introduction to the 2005 MVVA Park Master Plan, landscape historian Ethan Carr noted:

> *New York's parks launched a national tradition of park making because they fused social functionality with the emotional impact of landscape aesthetics. Central and Prospect Parks succeeded as truly public art: sequences of landscape scenes that also served vital needs for fresh air, open spaces to play, lakes for boating and skating, and paths for walking and riding. This required a kind of alchemy that took the ordinary features of a site—rock outcrops, hydrologic patterns, topography—and transformed them into nobler and more powerful versions of themselves.*

Experiencing Carr's insights in person, seeing how these two parks still capture one's imagination, was well worth the short subway rides from our office on 17th Street in Manhattan and later from Court Street in Brooklyn Heights. But to find that level of experiential power derived from postindustrial sites, we looked farther afield. Seattle's Gas Works Park, by the landscape architect Rich Haag, provided an inspiring comparison. We appreciated that Haag neither tamed nor ignored the ruins of rusting metal pipes and tanks, and how his oversized, abstract landforms encourage visitors to consider what strange history brought them together with the industrial remnants. We also discussed Jean-Charles Adolphe Alphand's Parc des Buttes-Chaumont in Paris. Opened in 1867, the sixty-one-acre park bounded by dense city blocks has a far different relationship to its context than our park does, but its aspirations were similar. A spent gypsum and limestone quarry in a then-marginal industrial area became one of Paris's most dramatic public parks, with dizzying paths swooping up and down wild constructed landforms while threading through thickets of trees and plants. It was crucial for us to identify which found features of the former docklands held significance and then determine how they might contribute to the material character of a new landscape.

Cultivating the image of a new park, in the studio and out in the world, was the only way to forge a shared understanding of the site's potential. At the core of this creative exercise was imagining a new relationship with the water. As obvious as it seems now, encouraging visitors to engage with the East River—considered by many two decades ago to be irredeemably industrial—took time and effort. Toward the end of a contentious public meeting in 1999, one older local resident, Mickey Murphy, walked to the front of the crowd and said, "I am retired. I live on a fixed income. I can no longer go to the country for vacations. I want to be able to go down to the East River at night, put my feet in the water, and see the reflection of the moon." From that evening forward, we no longer thought of the East River as a boundary but as an inviting and boundless extension of the park.

PROGRAM FOLLOWS PLACE

Through dozens of other public meetings, it became clear that the park would need an expansive range of recreational spaces to make up for what Brooklyn lacked. Athletic fields, kayak launches, and grilling areas were just a few of the must-haves. But offering specific activities without creating a sense of place would limit the ways people could enjoy the new park. When discussing program in isolation from landscape experience, it is easy to get mired in arguments about what to include or leave out. But when people spoke generally about how they wanted to feel in the park, we recognized the way forward to a more productive conversation. Instead of presenting a list of the activities different areas would be used for, we developed a set of terms to capture how those areas could feel. Words like "civic," "wild," and "boundless" were paired with universally recognizable images of busy plazas, meadows, and unending vistas. This helped us break the stalemate of competing programs and move on to the many worlds we were building in the park.

VISUALIZING

As excitement for the park grew, few people had a sense of the scale of the piers and the parking lots, or the vastness of the surrounding waters when seen up close. For the community to comfortably engage in conversations about the park's design, we needed to give them a more intuitive understanding of the site in all its immensity and complexity. Even the most precise drawings can fail to communicate the experiential qualities of open space, elevation change, materials, or vegetation. So we presented large, scrappy physical models designed to show three-dimensional space as clearly as possible. The bigger the model, the more easily people could imagine themselves within the future park. The final site model was fifty feet long, which required a storefront below our office to display for the crowds who attended public meetings.

Another way of establishing a shared vision for the design was through eye-level perspective sketches. Our renderings were inspired by planning drawings for Christo and Jeanne-Claude's sculptural installations. We borrowed the gestural, hand-drawn character of those sketches to give our images a work-in-progress feel. Layering loose pencil drawing over digital photo-collage of the existing site, our "scribble technique" was meant to start rather than end conversations. One drawing shows a boardwalk reaching across a sprawling new salt marsh in a cove just north of the Brooklyn Bridge. No such place exists today, yet the sketch helped build consensus for the bold idea of constructed salt marshes, which were eventually realized in other parts of the postindustrial shoreline.

ORGANIZING THE PARK

Every MVVA project involves a leap of imagination, but we also pride ourselves on our practicality. Designing this park helped us appreciate these mindsets as two sides of the same coin. Upon taking over as the leader of the planning and design team, we faced a rapidly changing urban context, including future development at the edge of the site, and layers of port infrastructure in every stage of disrepair. We knew the design process would span decades and entail countless incremental decisions, so we needed practical skill to think on our feet. And just as much, we needed bold, clear concepts and principles to stay focused on the big picture as the details evolved.

One big, overarching strategy came from the often-voiced concern that some of the piers were so far from entrances that no regular visitors would venture out to them.

We realized that sports courts and recreational fields were in such high demand they would generate constant use no matter where they were located. By situating them out on the harder to reach piers, life and energy could be drawn across the full expanse of the site. And with the popular athletic facilities concentrated on Piers 2 and 5, the other piers could provide more immersive landscape experiences, offering tranquil moments that are rare in the city and a chance to wander in the safety of a well-used park.

The park's circulation emerged from the limitations of its urban context. Between Atlantic Avenue and the Brooklyn Bridge, the site was largely cut off from adjacent neighborhoods by the Brooklyn-Queens Expressway and a bluff almost one hundred feet tall, compounding the feeling of isolation. To give the long and narrow southern half of the park continuity, a thirty-foot-wide walking and cycling path called the Greenway threads together piers, shorelines, and uplands. A series of narrower secondary paths lead off this main promenade—some extending out onto the wide open margins of the piers, and others ascending into the leafy uplands. With twenty feet for pedestrians and another ten for bikers, the Greenway is a generously sized corridor that connects nearly a mile of the park. The bike paths became part of the Brooklyn Waterfront Greenway, a protected lane that stretches from Red Hook to DUMBO. North of the Brooklyn Bridge, where the site and city are at the same elevation and free from the expressway, the park's bike paths are tied directly into the street network. Despite its once limited connections,the park has become inseparable from the city.

Found conditions across the site further informed the organization of the park. The project engineers assessed the structural capacity of existing marine structures to develop what we affectionately called the "Stoplight Plan." Using a color-coded system, the plan identified the condition of each built feature along the shoreline: green for those in good enough shape to preserve, red for those that had deteriorated beyond repair, and yellow for those that could go either way. Among the most significant findings was the discovery that twenty acres of the site were supported by 12,000 wooden piles that in many cases would require costly repairs. To make the most of a limited budget, MVVA chose to locate lightweight programs like athletic fields, courts, lawns, and meadows on the pile-supported piers while reserving the solid ground for new buildings and heavier landforms. We applied a similar principle when carving away the shoreline to invite bits of the East River into the site, wherever collapsing structures would otherwise need to be replaced. This not only saved money but allowed us to make evident in the new park the site's industrial past.

A MATERIAL LANGUAGE

The history of Brooklyn's waterfront is a cultural touchstone that we were determined to reveal in the landscape. Frugality again proved to be the best guide. Reusing structures and salvaged materials created an authentic visual identity while reducing costs, and our designs prioritized durability over elements that would require more frequent repair. As a result, the site's history is embedded in the material language of the park, through preserved structures and a gritty pragmatism that would be familiar to a Brooklyn dockmaster of generations past.

We preserved portions of the existing sheds that jut out into the river—including steel beams, columns, and low perimeter concrete foundations—on four of the park's five piers. These were refurbished and reinforced, creating consistent envelopes for various park experiences. The frames, a trace of the industrial past, were painted the same blue that the Port Authority had used on the sheds decades earlier. Some were retrofitted as armatures for lighting, shade structures, or netting to contain errant volleyballs. On Pier 2, even more of the original structure was retained to create a sports pavilion by Maryann Thompson Architects and Easton Architects. Sections of steel trusses, purlins, and bracing were adapted to allow large portions of the corrugated roof to be reinstalled for sun and rain protection. The sides were opened all around to invite onlookers, to let in cooling summer breezes, and to accentuate river views.

Every picnic table, most benches, and even some play equipment were built from remilled longleaf yellow pine timbers that were salvaged from the National Cold Storage warehouse. Virtually extinct as a commercial lumber source after centuries of overharvesting, this southern tree was long prized for exceptional rot resistance and strength.

The site as we found it could have easily come across as a list of problems to solve. But we understood it in a more positive light.

Longleaf yellow pine once underpinned the city's subway tracks. Today it clads the pavilion on Pier 6 as well as the entryway that passes through the 1 Hotel Brooklyn Bridge, serving as a design cue that guides visitors throughout the park. The same qualities that appealed to builders of the Brooklyn port make longleaf pine an exceptional material for park furnishings. The choice gives users a substantive connection to history, not just a symbolic one.

Facing a wide range of landscape conditions at the park's inception and anticipating more variables as it matured, we selected unfussy surface materials and park fixtures that would have universal appeal and be easily adapted to different circumstances. The paths are largely asphalt with poured concrete inserts for the mounted furnishings, and the broad perimeter walks at the water's edge of all five piers are the same concrete found there in 1999. Some materials were chosen that could serve dual purposes. Stone riprap is used to stabilize the shoreline, but it is equally adept at retaining upland slopes, holding soil for plants, acting as informal seating, and doubling as guardrails at path edges. The riprap boulders are even used in the uplands to provide a rugged, irregular counterpoint to the large, square-cut reclaimed granite blocks that were salvaged from the reconstruction of the nearby Roosevelt Island and Willis Avenue Bridges.

The miles of wire fences that protect plant beds are made from regionally harvested black locust posts, with a rustic, approachable look and the same kind of adaptability as farm fencing. The primary light fixtures are attached to forty-foot-tall wooden poles across the site, which would be as appropriate for a pastoral road as they are for the postindustrial waterfront. Commonly employed in marine facilities, galvanized steel was chosen for the railings that trace the perimeter of all five of the piers. Its endurance in coastal environments remains as relevant today as it was when the pier sheds were first built.

THE OPTIMISM OF BROOKLYN BRIDGE PARK

By the time Hurricane Sandy struck in the fall of 2012, flooding the site with vast quantities of saltwater, Pier 1 had been complete for a few years, improvements to the Empire Fulton Ferry section of the park were finished, Piers 5 and 6 were partially built, and the Greenway ran about half of its eventual length. That summer, a pop-up pool and temporary beach near Pier 2 had built excitement for the park spaces to come. The storm brought another image of the park, one where Jean Nouvel's glowing glass box around Jane's Carousel, illuminated by generators and surrounded by floodwaters, was for a moment one of the only lights on the East River shoreline.

As the water receded, we discovered the park was still very much alive. Our emphasis on pragmatic, economical design resulted in a living infrastructure that thrives amid the effects of climate change. At MVVA we never think of "resilience" as an isolated consideration to fold into our design strategies. We believe environmentally sound and durable landscapes are a natural outcome of rigorous, all-encompassing, forward-looking design.

During times of distress, the value of good parks is revealed to us in ways that we may otherwise take for granted. We anticipated that the landscape would have to prove itself in increasingly likely flood events. We were less prepared for the global outpouring of renewed support for public spaces resulting from the COVID-19 pandemic. Our greatest hope for Brooklyn Bridge Park, however, was that it would have a place in the heart of everyday life in the city. And for parks to work effectively in times of upheaval—as they are being asked to do more than ever before—they need continuous investment from institutions and local communities during life's quieter moments. Parks need to be supported like any other essential infrastructure and understood as living and evolving environments. We strove for Brooklyn Bridge Park to embody this principle. Designing it to function was no more essential than designing it as a place to be loved.

LANDSCAPE IMAGINATION

The industrial seawalls were failing, which created opportunities for a more functional, accessible condition between land and water.

ABOVE *An early MVVA sketch of a salt marsh, never realized, in front of the Empire Stores, 1999.*

ABOVE *Planning principles from the pages of the* 2005 Master Plan: A Framework for Design.

RIGHT *Meetings between MVVA, collaborators, and public stakeholders held around site models, 2003–5.*

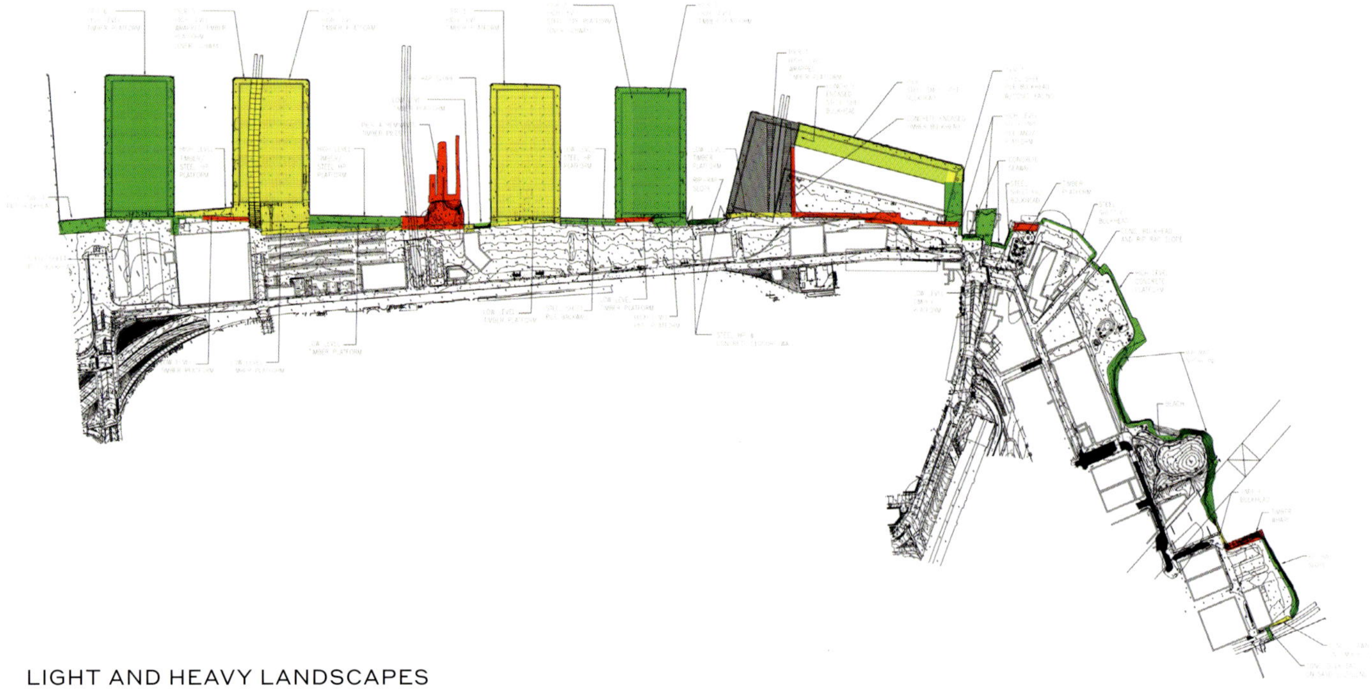

LIGHT AND HEAVY LANDSCAPES

Decisions about where to add topography and where to retain the existing flat surface were guided by the load-bearing capacity of each area.

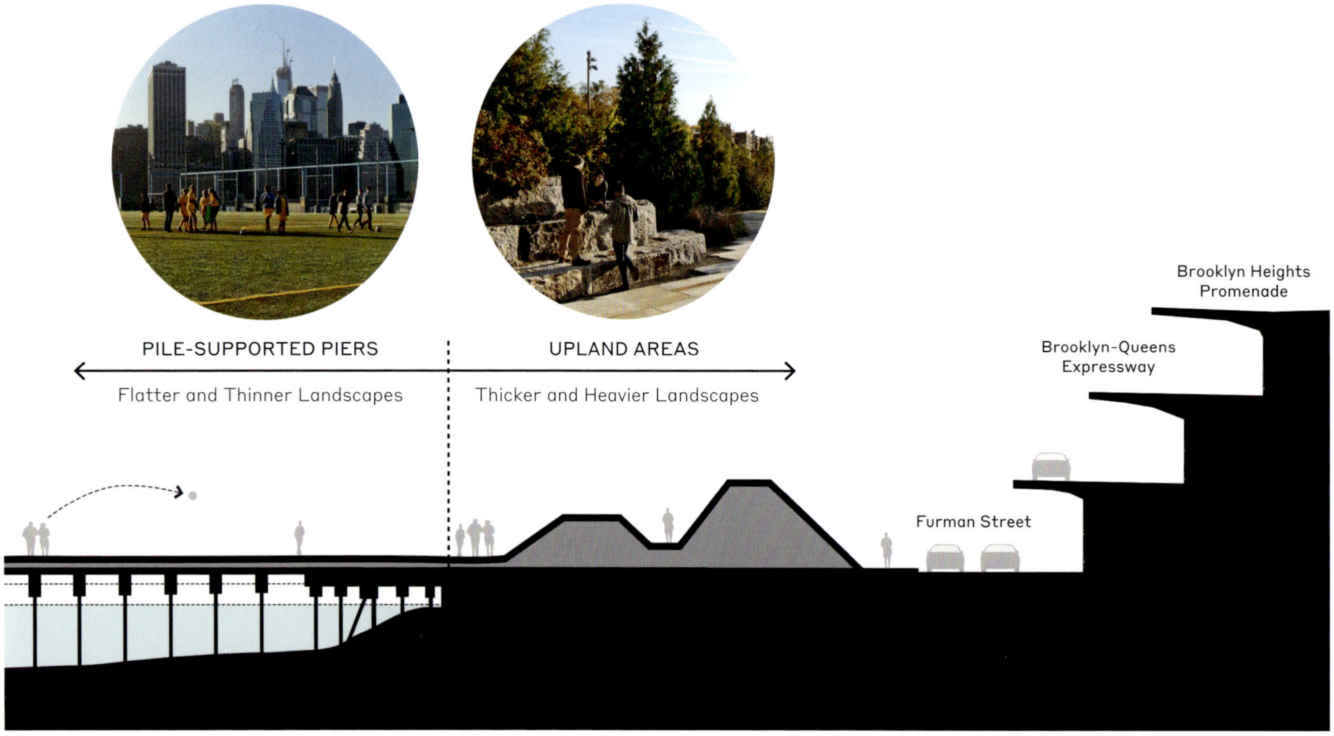

At Pier 5, a sound berm was constructed on the solid ground of the uplands, while a soccer field occupies the level pier above the water.

PIER 6
PIER 5
PIER 3
PIER 2
PIER 4
BROOKLYN - QUEENS EXPRESSWAY (BQE)
Brooklyn Heights Promenade
COLUMBIA HEIGHTS
BQE
Adam Yauch Park
ATLANTIC AVENUE
STATE STREET
JORALEMON STREET
REMSEN STREET
MONTAGUE STREET
PIERREPONT STREET
HICKS STREET
BROOKLYN HEIGHTS
CLARK STREET
HENRY STREET
CLINTON STREET
0
250
500
1000

BROOKLYN BRIDGE
PIER 1
EMILY ROEBLING PLAZA
EMPIRE FULTON FERRY
MAX FAMILY GARDEN
JANE'S CAROUSEL
FURMAN STREET
Squibb Park
Hillside Dog Park
OLD FULTON STREET
WATER STREET
BQE
MIDDAGH STREET
MAIN STREET
MANHATTAN BRIDGE
MAIN STREET
DUMBO
WASHINGTON STREET
JOHN STREET
PLYMOUTH STREET
JOHN STREET
CADMAN PLAZA WEST
Cadman Plaza

PREVIOUS PAGES *From above, the lush planting of Piers 1, 3, and 6 contrasts with the playing fields of Piers 2 and 5.*

LEFT *The Pier 5 soccer field is used by adult and youth teams from all five boroughs.*

MATERIAL LANGUAGE

Drawing on the everyday, unfussy materials of the former port makes the park more welcoming and helps the landscape stand up to heavy use. The landscape includes adapted pier structures, reclaimed materials from the site, and salvaged resources from city infrastructure.

ON-SITE HARVEST

Dismantling the National Cold Storage warehouse produced nearly one million board feet of heritage longleaf yellow pine lumber, which was remilled and used for benches, picnic tables, and playground equipment.

ON-SITE REUSE

Portions of the concrete deck and piles at Pier 3 were removed to reveal the water's edge and relocated to the uplands of Pier 2, where they are now sculptural features in a water playground.

CITY SALVAGE

Massive amounts of granite were salvaged from the reconstruction of two New York City bridges and incorporated as embankments and seating. NYC Parks Commissioner Adrian Benepe told MVVA about the availability of the granite, saying "you'll find a use for it." The stone cladding later inspired the Granite Prospect.

457
WATCHTOWER

MEETING THE CITY

MEETING THE CITY

An immediate challenge of planning Brooklyn Bridge Park was forming identifiable and inviting entrances into the landscape. Deeply removed from the existing city fabric, the park would be encountered naturally only if we created conditions for that to happen. The triple cantilever of the Brooklyn-Queens Expressway (BQE), hovering above the equally gritty Furman Street, severs the Brooklyn Heights neighborhood from the waterfront and leaves only three possible entry points along nearly a mile of the site's urban boundary. Between Piers 1 and 2, the BQE curves inland, allowing for easier access to the northern waterfront. But even there, largely vacant nineteenth-century warehouses crowded the edge of the site, which made the existing entryways anything but welcoming.

The MVVA team explored many ideas for new ways to reach the park. The only feasible option turned out to be a new bridge that leaps over Furman Street from the existing twentieth-century Squibb Park in Brooklyn Heights and touches down on Pier 1. Otherwise, the park design needed to make the most of the already existing entry points, all of which came with the distinctly unwelcoming character of a transition into an industrial zone. The sense of vibrant city life quickly fell away as we crossed into the vacant site. The solution to make the entries feel inviting was to bring the life of the city into the park by concentrating human activity at each threshold. The 2005 Master Plan described the result of these entrances as "urban junctions" that leverage every possible means of hardwiring connections between the park and its neighborhoods.

The essential criteria for urban junctions included generously sized gathering areas with ample seating and lighting, broad entry paths with clear sight lines deep into the park, and distinctive plantings that signal the presence of the park from the outside and encourage movement. High-use park features are clustered at the junctions to attract consistent attention from both park visitors and local residents: things like playgrounds, dog runs, passenger ferry landings, bikeway connections, food concessions, and restrooms. By providing amenities where they were needed most, this strategy draws visitors farther into the park.

It is an age-old practice to locate things that busy urbanites need from a park along the site's exterior, but MVVA took the idea a step further, shaping not just the park, but how the city gathers at its edges. By integrating apartments, a hotel, and commercial space with the landscape design, the urban junctions sustain the park socially as well as financially.

FINANCIAL SELF-SUFFICIENCY AND INVITING THE CITY IN

Beyond improving access to the park, the urban junctions satisfied an important condition for the park's financial infrastructure: Brooklyn Bridge Park is required to generate revenue on-site in order to fund its maintenance and operations. The city and state agencies issued this mandate in exchange for guaranteeing public funding for the park's construction and to avoid a repeat of funding gaps that led to the decline of care for New York City public landscapes in the 1970s. To cover long-term costs, 10 percent of the land at the city-facing edge of the park was set aside to provide financing through tenants' ground leases and payments to the Brooklyn Bridge Park Corporation in lieu of taxes. MVVA was charged with integrating this money-making real estate into the landscape, perhaps the most controversial element of the project.

Rather than treat the income-generating parcels as losses for the park, we saw them as an opportunity to create better connections to the city. The plan incorporated the new buildings into the major park entrances at the ends of Atlantic Avenue and Old Fulton Street, placing residents and workers exactly where around-the-clock activity would be a needed asset. The presence of people who treat the park as a backyard extends the hours of use earlier into the morning, later into the evening, and through the winter, something that would not be possible if development had been prevented from crossing the old dividing line between neighborhood and waterfront.

Stitching city and park together required tricky coordination between the design of the public landscape and that of the adjacent development. MVVA helped select development proposals that both generated the essential financial returns to sustain the park and fostered ground-level activity—all while preserving as much parkland as possible. Over the course of planning and design, the park's footprint stretched to incorporate Pier 6 to the south and three acres at John Street to the north. These parcels linked the park to additional parts of Brooklyn through their proximity to Atlantic Avenue at one end and Jay Street at the other, both major thoroughfares that span much of the borough. The length of the park site forged connections across more neighborhoods, but it also resulted in a wide variety of challenges and opportunities at each boundary with the city.

We recognized three basic edge conditions along the interior of the park site: the boundary south of the Brooklyn Bridge, characterized by extreme elevation change, with the BQE separating the waterfront from Brooklyn Heights; the densely packed blocks of DUMBO (Down Under the Manhattan Bridge Overpass), where the warehouse district had become a hotbed of stores, galleries, and tourist sites; and the area directly under the Brooklyn Bridge, a complex, historically loaded intersection between the two segments of the park. Addressing these conditions, while preserving continuity in the park experience, ultimately gave each junction its character.

FROM THE HEIGHTS TO THE WATERFRONT

From the park's southernmost boundary at Atlantic Avenue to Old Fulton Street at the north, the historic neighborhood of Brooklyn Heights sits on a bluff high above the waterfront. For most of that distance, two tiers of the BQE are embedded in the hillside directly below the cantilevered Brooklyn Heights Promenade, a destination with dramatic views but no access to the water. For nearly a mile between Atlantic and Old Fulton, only Joralemon Street—a shady, one-way, cobblestone lane of townhouses and apartment buildings—crosses beneath the forbidding expressway to reach the interior of the park.

A main artery to the historic waterfront that spans the entire borough, Atlantic Avenue possessed the ample width and connectivity needed to be a major gateway into the park. Given the historical priority of motor vehicles and port services, however, the intersection of Atlantic Avenue at the park boundary was not designed to feel comfortable for people on foot. We could not change the street layout, so we had to help visitors understand how to enter the park and to feel safe doing so.

Today, an existing sidewalk passing through an underpass of the BQE leads visitors to within sight of the main entry to the Pier 6 uplands. Aligned with the city sidewalk, the park walkway is framed by a dense allée of common privet, untrimmed and reaching twenty feet high. A playfully oversized version of a common residential screen, the billowing privet hedge overhangs rows of benches on both sides to form a shaded archway that is extremely popular on hot summer days. There is a calculated gentle rise and fall in the path as it gradually curves away from the city grid that creates a subtle transition away from the Atlantic Avenue streetscape while setting up a clear sightline deep into the park. Two new residential buildings that fund park maintenance, a public dog run, and two acres of planted play spaces in the uplands of Pier 6 ensure that the terminus of Atlantic Avenue is alive with activity day and night.

To reinforce a sense of arrival in the park interior, we worked with Maryann Thompson Architects to transform a concrete headhouse at the base of Pier 6 into a multi-use park building. The Warming Hut offers public restrooms and seasonal dining areas, both indoors and out. The roof, made accessible by a ramp clad in reclaimed long-leaf yellow pine, is home to a popular pizza restaurant and gives visitors an elevated view of the harbor, the Manhattan skyline, and the sand volleyball courts next door.

Although more comfortable to walk than Atlantic Avenue, Joralemon Street still faced a stark difference in character at the park edge, with the BQE and Furman Street marking a hard line from neighborhood to industrial zone. Today, active storefronts draw day-to-day city life across this once-harsh boundary. A small street extending from Joralemon into the park—renamed Brooklyn Bridge Park Drive—is flanked by two historic prewar warehouses that were transformed into buildings that now support the park. The H-shaped 360 Furman Street became the One Brooklyn Bridge Park condo block, while a smaller brick warehouse was converted into the offices of the Brooklyn Bridge Park Conservancy. From there, Brooklyn Bridge Park Drive leads to Pier 5, where the movement of soccer players on the fields and families milling about the shaded picnic tables creates a welcoming scene. The ground-level cafés and shops make the Greenway, the main park promenade, feel more like a lively neighborhood street. An ice cream shop just outside the soccer field complements the small children's play area, public grills, and the seating areas of the Picnic Peninsula.

Farther north, the boundary between Pier 1 and Brooklyn Heights is more removed from the oppressive traffic noise of the BQE, but it came with an unwelcoming mix of retaining walls and loading docks on Furman Street, which was historically the back-of-house side of the Old Fulton business district. Here, the 1 Hotel Brooklyn Bridge and Pierhouse residential development, designed by INC Architecture and Marvel, create a welcoming threshold. A break in the building complex doubles as a drop-off circle and pedestrian entry point from Furman Street, and other gaps in the building provide access to the park, including a three-story gateway leading through a planted courtyard to the edge of Pier 1—a nexus of activity populated by hotel guests, coffee shop patrons, and parkgoers.

URBAN INTENSITY IN DUMBO

The Empire Stores, a large former warehouse renovated by Studio V and S9 Architecture, borders the park at its narrowest point in the DUMBO neighborhood. Barely thirty feet separates the facade of the building from the riprap embankment and sand beach at the river's edge. Along the length of DUMBO, the city fabric encroaches so severely on the shoreline that there was no choice but to incorporate highly urban conditions into the park experience. At the Empire Stores, reactivating the ground floor with retail and restaurant activity spilling through the stately brick archways to outdoor cafés allowed the thin strip of park to become a cosmopolitan promenade. At the same time, an open-air courtyard spanning all four stories of the original structure allows the public park use to reach into the envelope of the building and even up to the roof.

To the west of the Empire Stores facade, another adapted historic building houses the St. Ann's Warehouse theater company and the Max Family Garden, which occupies a two-story triangular brick shell. The intermingling of city and park life is particularly pronounced along this stretch, where Empire Stores and St. Ann's Warehouse each have highly visible frontage on the park side as well as on DUMBO's high-activity Water Street.

A block east, Washington Street is a well-known approach to the waterfront, endlessly photographed for its cobblestone paving, SoHo-style brick loft buildings, and distinctive framing of the Empire State Building through the tower of the Manhattan Bridge. It becomes one of the park's main urban junctions at the Main Street section, where a former masonry maintenance building was transformed by Architecture Research Office (ARO) into a park operations facility, welcome center, and the Brooklyn Bridge Park Conservancy's Environmental Education Center. On one side of the Washington Street entrance is a beloved maritime-themed play area preserved from an earlier iteration of Main Street Park designed by NYC Parks and Recreation. On the other, directly below the Manhattan Bridge, are The Cliffs at DUMBO climbing facility and Main Street Dog Run, which now sits across the street from a new branch of the Brooklyn Public Library. Clustering active uses like these in a way that complements the growing cultural life of the neighborhood is essential for creating a sense of cohesion where the park might have otherwise been the most fragmented.

UNDER THE BROOKLYN BRIDGE

While developing the Master Plan, we were conscious of the complexity and historical significance of the entrance at the end of Old Fulton Street. In a sense, this was the only urban junction that already existed, with the cherished Bargemusic concert venue, popular ice cream and pizza shops, the venerable River Café, and several blocks of comfortably scaled streetscape. We were happy to let this urban ensemble continue to develop on its own, but creating clarity in the unresolved space directly under the bridge was essential to building continuity across the park's north and south sections. In 2008, the demolition of the Purchase Building located directly beneath the bridge cleared the way for continuous foot traffic from one end of the park to the other.

Emily Warren Roebling Plaza, built in 2021 on the site of the Purchase Building, became the last major space to be completed in Brooklyn Bridge Park. It now acts as a civic space that gracefully ties together the two park sections and two neighborhoods. The purpose of the plaza was not

Deeply removed from the existing city fabric, the park would only be encountered naturally if we created conditions for that to happen.

to introduce new activity but to absorb the spillover of city life from an existing hub of city life. Planted mounds at the street edge loosely define a central gathering area without strictly delineating a prescribed space, which might have felt empty when unoccupied. And when the city is most alive, this largest clear paved area in the park can support events at a scale that fits its monumental backdrop.

PIER 1 ENTRY REDESIGN

Long after the urban junctions were formulated, MVVA and the Brooklyn Bridge Park Corporation are still experimenting with ways to make them better. The main entrance to Pier 1, completed in 2010, was not built to accommodate the millions of visitors who come to the park each year.

An opportunity to redesign the entry emerged when the nearby service drive for the 1 Hotel Brooklyn Bridge became obsolete. Since 2021, MVVA has been working with Tod Williams Billie Tsien Architects to clarify the entry and routes to the park interior, welcome the ever-increasing number of visitors, and activate the threshold with a café and restrooms. The reimagined Pier 1 entry features a paved bosque and an open pavilion designed by TWBTA. A mixed grove of high-limbed honey locust and ginkgo trees will allow views into the Bridge View Lawn while providing dappled shade and a beacon of green from the street.

Designing a park that is so embedded in its context comes with a special challenge and a gift: the park is forever accountable to—and conjoined with—the city beyond its boundaries. The urban junctions have taken on lives of their own, evolving alongside the streets, institutions, and people who define their utility and character. So, instead of will people come, the question now is how can the park better serve even more visitors in a changing urban environment. A place like Brooklyn Bridge Park does not form just one relationship with its city. Park and city meet, over and over again, every day.

The Squibb Bridge crosses over Furman Street and leads visitors directly into the park near Pier 1.

FDR Drive
Brooklyn Bridge (NYC DOT)
South Street Seaport
EAST RIVER
Empire Fulton Ferry Landing State Park
Main Street Park
Water Taxi
CON ED
The Cove
Jay Street
Fulton Ferry Landing
York Street Station
Manhattan Bridge (NYC DOT)
DUMBO
Old Fulton Street
Brooklyn-Queens Expressway
Pier 1
Squibb Park & Bridge
Pier 2
High Street Station
Adams Street
Cadman Plaza
Clark Street Station
Pier 3
Furman Street
Brooklyn Heights Promenade & Pierrepont Playground
BROOKLYN HEIGHTS
Pier 4
Marina
Montague Street
DOWNTOWN BROOKLYN
Court Street Station
Pier 5
Brooklyn-Queens Expressway
Borough Hall
Joralemon Street
Borough Hall Station
Pier 6 (City of New York)
Water
Atlantic Avenue

URBAN JUNCTIONS

To overcome the isolation of the site, we positioned park entrances at the ends of major streets and clustered development parcels as well as everyday park programs around them.

FROM THE HEIGHTS TO THE WATERFRONT

Privet hedges overhang rows of benches on both sides of the Atlantic Avenue entrance to form a shaded allée that is popular on hot summer days. Gaps in the rows of privet lead to five unique playgrounds.

LEFT *The Atlantic Avenue entrance to the uplands of Pier 6.*

RIGHT *The garden rooms of Swing Valley at Pier 6.*

PLACES FOR PLAY

While the MVVA team was planning the park in the early 2000s, countless families with young children were moving to nearby neighborhoods. Both newcomers and longtime residents made it clear that they wanted a park their kids would love. We also recognized that the park would be a much-needed resource for many of Brooklyn's 2.4-million residents, including children from neighborhoods where opportunities for play were limited.

At the time, we were expanding our approach to designing areas for play, adding elements that were not only physically challenging but also socially and cognitively engaging. Our goal was to provide children with various ways to learn about themselves and the world they live in. In 2004, while our work was underway at Brooklyn Bridge Park, we completed Lower Manhattan's Teardrop Park, a place shaped entirely for kids and inspired by nature. Bounded by tall buildings, the 1.8-acre void in Battery Park City became a microcosm of the Catskill Mountains. The dramatic topography, natural stone, and dense plantings both defined play spaces and served as informal play elements themselves, with rock faces to climb and miniature woodlands to explore. Teardrop Park marked the first time MVVA worked with play experts Robin Moore and Nilda Cosco of the Natural Learning Initiative at North Carolina State University's College of Design, who both inspired a revolution in the way we thought about nature play.

With Brooklyn Bridge Park's eighty-five acres, we had the chance to build on Robin and Nilda's ideas under more varied conditions and on a bigger scale. Our philosophy was that great play spaces need a strong connection to nature, and that inclusive design for children can draw on the complexity of natural environments. Integrating play into a landscape provides a sense of belonging for children with all kinds of needs, which a playground defined by freestanding, standardized equipment cannot offer. Together, our team began identifying a spectrum of spaces to serve different personalities and play styles—from central, highly visible areas for children who want to direct their energy outward to sheltered nooks at the edges for those seeking calm or interaction with smaller groups. Robin and Nilda told us that a successful play environment requires caregivers' cooperation, too. To create somewhere that adults would want to return to, we provided a range of sitting options that allow them to be as involved or as hands-off as they please. When these spaces are part of a holistic landscape, they can overlap and diverge seamlessly so that participants can find their comfort level and move in and out of each zone with ease.

These characteristics are especially important for welcoming children with cognitive and physical abilities not served by traditional playgrounds. The park's play spaces offer opportunities for climbing, crawling, jumping, and running, sometimes at very high levels of physical challenge, but they are also designed so that anyone can simply enjoy the fundamental experience of moving through and sensing the landscape. We found that richly planted spaces are an effective component of "universally accessible" play, as plants and other natural materials provide sensory input compatible with various modes of movement, perception, and processing.

We sourced play equipment from innovative vendors and designers to achieve the best and most practical fit for the landscape. One of our most productive partnerships is with the German company Richter Spielgeräte, which has been creating complex play equipment with an emphasis on natural materials since the 1960s, when European ideas of independent adventure play first caught the attention of American city planners and designers.

Play was an essential piece of our effort to animate park thresholds with regular activity, so playgrounds were located near the entrances at Main Street, Pier 1, Pier 5, and Pier 6. In addition to enlivening the moment of arrival, placing play near these edges has practical benefits for children's caregivers, who may have limited time and ability to move farther into the park. As the park design evolved, we identified other opportunities for a more informal kind of play and created areas that encourage exploration and problem-solving while blurring the lines between designated program and the experience of the landscape as a whole.

Nestled in a horseshoe bend of the path in the Pier 2 uplands, the integrated water play area is a serendipitous encounter for most visitors. For determined kids who find the foot-operated buttons, water jets turn a series of monoliths into fountains. The objects whimsically arrayed around the ground-mounted water spouts consist of rough-cut concrete pile fragments that were sawn away

Whatever else a park may be, it needs to be a place where people move their bodies and exercise their imaginations for no other purpose than the pleasure of it.

from the deck of Pier 3. When the water is turned off in colder months, the piles become quiet monuments to the site's industrial past.

Without instructions or prescribed interactions, these play elements reward experimentation, teamwork, and return visits to discover their full potential. In turn, the hybrid play-gardens become social ecosystems, as kids of different ages and backgrounds share discoveries with their friends and caregivers. These spaces help open up the rest of the park to interpretation, giving users license to treat other park elements as playful: the stacked stone and the open window portals in the Max Family Garden, the sculpted riprap along the water's edge, or the many grassy slopes that invite running, rolling, and sledding.

It is no accident that the park's largest play area on the uplands of Pier 6 is located at what was once the most unappealing entrance to the site. Beset by highway noise and acres of treeless concrete, the intersection of the park with Atlantic Avenue needed to deliver a powerful energy and sense of welcome to counteract the harshness of its setting—much as the whole park does for the waterfront.

The 1.6-acre Pier 6 play area embodies all our core principles of play design, activating a major entrance while also offering quiet moments of discovery. Five separate garden-like rooms are embedded in a lushly planted environment and carved into a robust topography that deflects city noise and blocks views to the still active port infrastructure nearby. In Swing Valley, bowls in the rubberized surface are as much a part of the experience as the swings. This fluid topography is accentuated all summer by bobbing globes of hydrangea and smoke bush flowers. At Slide Mountain, the same cresting landform that buffers the park from the street supports a gently sloping wheelchair-accessible path to the top of two super-sized slides. Both the path and a rock scramble shortcut are integrated with dense stands of privet and bamboo—jungle-like at the scale of a child—so that each option is an exciting and unexpected experience of landscape.

The child-focused planting palette at Pier 6 shifts in character with each space: Slide Mountain's bamboo gives way to katsuras, catalpas, and sumac, whose expressive flowers and foliage suggest an almost tropical setting. The trees encircle Sandbox Village and the Water Lab, the latter devoted to adventurous ways for children to feel, touch, and manipulate water. Designed for experimental, creative play, these areas owe something to the philosophy of unrestricted exploration that inspired New York's mid-century "adventure" playgrounds. But instead of sparking the imagination with geometric forms, they offer the infinite variability of sand, water, and plants, along with a series of objects ranging from recognizable animal forms to abstract kinetic machines. Around the corner, a garden of native plants called the Exploratory Marsh, with a small path of long, rough-sawn log planks, has become a beloved secret space for children.

Tall, untrimmed privet hedges enclose the Pier 6 play areas on either side of a broad walkway that leads visitors into the park. There is a reason why even a parkgoer with no particular interest in play is routed past the filtered sounds of laughter, pounding feet, and splashing water, along with the occasional glimpse of kids darting around in excitement. The experience of joy, whether as a child, as a caregiver, or even as a passerby, is central to how the Pier 6 uplands transition from the hard-edged world of the city to the embrace of the park.

Whatever else a park may be, it needs to be a place where people move their bodies and exercise their imaginations for no other purpose than the pleasure of it. Yet when we talk about play, of course, the focus is on children. At Brooklyn Bridge Park, we thought about how kids would engage the entire landscape. However, we learn something important by paying attention to places designed specifically for play. Lacking the inhibitions of adults, children clearly signal the success or failure of such places with their whole selves, and caregivers and onlookers take note. Many of us at MVVA have been fortunate to witness our children and grandchildren grow up in these miniature worlds. Kids embrace every element of the landscape—the sound of running water, the hide-and-seek game offered by a few trees, the license to run across wide-open lawns—in unexpected and delightful ways, forever teaching us lessons that improve how we design spaces for children.

LEFT *Slide Mountain at Pier 6.*

RIGHT AND FOLLOWING PAGES *Water Lab at Pier 6.*

DESTINATION
IMAGINATION
2014

PIER 6

LEFT *The Pebble Beach at Main Street Park beneath the Manhattan Bridge.*

FOLLOWING PAGES *The Empire Stores and Promenade at Main Street.*

west elm

THE MAX FAMILY GARDEN

The 8,000-square-foot Max Family Garden can feel like something made by a home gardener. Nothing planted or laid down in the hollow brick triangle on Brooklyn's DUMBO waterfront was too massive to have been carted in by hand. The largest of the salvaged curbstones arrived in a loader, but it's easy to imagine them all being muscled in on foot. The young trees, now thickened and sprawling, were slim enough when they arrived in the fall of 2015 to place without a machine. Of the thousands of bricks that make up the garden's paving—a more delicate surface than the concrete and asphalt found throughout the park—hundreds were positioned and repositioned individually as if by a dedicated amateur to form an abstract mosaic that construction drawings could only begin to suggest.

The sense of being handmade was not a design principle. None of us ever said, "Let's make it so anyone who visits the Max Family Garden could imagine making it themselves, so they will love it and feel at home there." In retrospect, it seems like an obvious place to start, as we wanted intimacy amid the monumental scale and energy of the garden's surroundings. But in our early designs, we had yet to realize the importance of cultivating this highly personal connection with visitors. Fortunately, our client helped steer us right.

In 2012, with Brooklyn Bridge Park under construction, MVVA was hired to create a landscape within the remnant brick walls of a nineteenth-century warehouse next to the park's Empire Fulton Ferry section. The St. Ann's Warehouse theater company had arranged a long-term lease to build its first permanent home on the site and had agreed to commission a garden they would turn over to the park upon completion. The theater would occupy a large, rectangular void in the old structure and leave the smaller, abutting triangle for the garden, which was expected to host spillover from events in addition to being a permanent feature of the park. This made St. Ann's founder and artistic director Susan Feldman our client, with input from Brooklyn Bridge Park staff and building designers Marvel Architects.

Our charge was to create a fitting threshold space for the theater that could double as a respite for the public from the busy neighborhood and park—a challenging mix of uses. The first design proposed an irregular field of "icebergs": large, angular slabs of exposed aggregate concrete pavement drifting together to form the main circulation. A thicket of gray birch would provide a canopy, and creeping perennials would fill the larger seams in the paving. We believed this repetitive but richly textured composition struck a balance between neutrality, for when the theater hosted an event, and visual drama, for when the garden was sparsely populated.

When we presented the idea to Susan, she gave us the go-ahead to begin design development. But soon after, she returned from a visit to the Burren, a coastal region of Ireland dominated by exposed rock and a diversity of flora, and she suggested a new vision of a treeless garden with a rocky tapestry of low, flowering plants, something strikingly different from the rest of Brooklyn Bridge Park. A creative professional well versed in collaboration across disciplines, Susan committed to supporting our original scheme if we thought it best, but she had clearly tapped into something more profound than what we had shown her.

We knew the reference to the Burren was more about a kind of energy than it was a suggestion to replicate an existing landscape. Though we loved the idea of a garden emerging from bare rock, tree canopy was still essential to moderate sunlight and weather and to engage the empty volume of the triangle. Susan's suggestion of a new direction got us to rethink sensory richness over adherence to our design concept. She wanted more personality, like the dramatic productions in her theater. Instead of a minimal composition of planting and pavement, we began to imagine a place with more twists and turns, like the deceptively complex Burren.

The angular slabs of concrete changed into a swarm of bricks. Where clear walking routes were needed from openings in the triangle to building entrances, a tight paving surface is made from a four-and-one-half brick pinwheel pattern. Along the path edges, the bricks spread and spin apart, dissolving completely into plants and soil or settling into a loose scatter strung out along secondary paths. In early design meetings, we had all agreed: no bricks—they'll clash with the existing structure. But as we began looking for ways to give our new scheme more depth, we inevitably gravitated toward a brick's small size, which can lend spaces a visual density to make them feel larger and more complex. We chose a mineral-brown, clay brick that

The long horizon is traded for a triangle of sky—an intimate alternative to the rest of the eighty-five-acre park.

contrasted well with the red walls and had a natural variation that would make the eye linger as if viewing across a greater distance than what is there.

In the initial design, we had originally planned to make a flagstone perimeter out of paving stone that had been salvaged from a nearby building. We instead inset the stone as pavers within the brick and heaved it up and out of the ground into gestural stacks that could serve as seating, informal play objects, and subtle delineations of space.

We also became more playful with the planting. We decided to make a plant-lover's garden first, a designer's garden second. To the birch trees we added sassafras and redbud, and over the course of several meetings with some of MVVA's greatest plant enthusiasts, the team expanded the list of shrubs and groundcovers to fill each season with a shifting ensemble of color, texture, and scent.

The process of reorienting the Max Family Garden towards simple pleasures and personal experience makes it exceptionally approachable for a public landscape. Being able to see clearly how it comes together—how a brick is placed, or a stone is stacked—and sensing the unpretentious joy in the planting, helps remove some of the distance that park visitors might feel between themselves and a professionally designed landscape.

A strength of Brooklyn Bridge Park is how it welcomes people to the awesomeness of their surroundings. The vast borrowed landscape of the river, bridges, and skyline, along with the energy of different kinds of activities happening all around, creates the feeling of communing with the life force of the city. Walking under the brick archways into the Max Family Garden, one crosses into a different kind of public space. The long horizon is traded for a triangle of sky. The cheerful din from Jane's Carousel 250 feet away still filters through the walls along with the churn of tourists on Water Street and the hum of traffic on the bridge above, but these things feel dreamlike and distant.

Instead of a mannered field of birch trees and poured concrete forms to remind everyone that this is a garden designed by landscape architects, we settled on a place with a this-one-here-that-one-there level of tenderness. Salvaged stone is informally piled in what feels like an invitation for visitors to decide for themselves how to use it, like the old window portals in the wall that offer the perfect perch or place to set down a cup of coffee. If the rest of Brooklyn Bridge Park makes you feel one with the grandeur of the city, this garden gives you your own piece of it. You belong there, but you also get to feel for a moment that the bricks, flowers, and stone—even the fragment of the bridge overhead—belong to you.

UNFLATTENING THE SITE

UNFLATTENING THE SITE

Arriving at the docklands for the first time, we encountered an expanse of level pavement that had outlived its purpose. It was more structure than landscape, with a hard and fixed edge built for industrial efficiency. As we explored the site, the panoramic views would quickly pass from spectacular to unrelenting. The harsh sunlight and the roar of traffic from the Brooklyn-Queens Expressway (BQE) made the site feel hostile. Making a comfortable and inviting park here would require dramatically reshaping the surface of the land.

Such was our thinking when, one day in 2005, the design team embarked on a boat tour of New York Harbor and the East River. We were looking for a range of natural and constructed waterfronts that could reveal something about our site and how to invent a new ecosystem. The varied shorelines of New York inspired us. We made notes of which structures, topographies, and ecologies seemed best adapted to storm surge flooding and other stresses of an urban coast, and we hoped to find clues about how to develop an authentic sense of place.

Heading upriver toward the Long Island Sound, we came upon a massive pile of gravel in the Brooklyn Navy Yard—an accidental landform that climbed to nearly thirty feet above the level docklands. One glance at this shift in elevation on the horizon convinced us that for the park to meet the ecological and experiential goals for the landscape, the topography needed to be anything but subtle.

BREAKING THE DATUM

The port's uniform parking lots and pier decks composed the datum against which every new contour and change in elevation was measured. To establish a dynamic experience that stretched the entire length of the park, we interrupted this baseline with a series of tall, earthen hills that diversify spaces and provided vantage points for taking in the view. Once these landforms were shaped, winding paths were laid out to rise and fall, shifting sight lines and amplifying the sensation of the landscape unfolding as one moves through it. Finally, we carved niches into the landscape to create spaces of intimacy and enclosure. Most of these features rise above the elevation of the existing piers and uplands. A few strategic excavations below grade have enabled new, dynamic shoreline conditions like the salt marshes and the wetlands. Selectively preserving limited stretches of level pavement has provided an essential counterpoint to the expressiveness of designed topography and planting. These areas help parkgoers appreciate the rhythmic shifts above and below the datum throughout the park.

HIGH POINTS

The landscape needed to offer unforgettable topographic encounters like the one we had observed at the Brooklyn Navy Yard, but the structural limitations of the piers and the site's narrow footprint limited the placement of such massive landforms—except on Pier 1. At nine acres, Pier 1 is 80 percent bigger than the other piers, and because it sits on earthen fill rather than on piles, it could bear the load of the materials required to build a twenty-foot-tall hill. And at 850 feet in length, Pier 1 could accommodate slopes that emulate the elongated form of a drumlin—a type of landform sculpted by the slow grinding movement of glaciers and their deposits. Although this kind of topography never existed there, it could have. It's a fantasy grounded in truthful possibilities.

There are two large, sloping lawns on either side of the drumlin—the Harbor View Lawn and the Bridge View Lawn—that frame vistas in opposite directions. Both are oval-shaped and bordered by plantings that embrace the lawns like forested ramparts. The one-acre Harbor View Lawn rolls down to form a grass amphitheater that faces the water, with the Statue of Liberty in the far distance. It ends at the pier's edge, meeting the existing concrete deck that serves as a wide perimeter walk. On the north end of the pier, the Bridge View Lawn slopes more gradually toward the Brooklyn and Manhattan Bridges. The hill's eastern side descends into a series of constructed wetlands before rising again to meet the Greenway, the park's main thoroughfare. The slopes of the landform create a varied distribution of moisture, sunlight, and wind exposure, yielding six distinct micro-environments, including sunlit lawns, miniature forests, and densely wooded hedgerows.

Because Pier 1 was the long-awaited first phase of the park, it had to set a high standard for future construction and illustrate myriad possibilities for the whole site. People would likely have flocked to the shoreline on opening day even if the park was left dead flat, appreciating the chance to stand by the water and take in the uninterrupted views previously obstructed by fences and derelict warehouses. But how long would they have stayed, and would they have returned, again and again? Unflattening Pier 1 allowed for a constantly shifting experience of the waterfront. We gave people reason to focus their attention landward, offering moments of surprise as they inevitably turn their gaze away from the water.

Farther north, at the base of the Manhattan Bridge and around a bend in the shoreline, is the Main Street section of the park. It came with an equally imposing context yet a much smaller footprint. In the early 2000s, the New York City Parks Department built a lawn and a cove-like beach with curved granite seating that steps into the tide and offers expansive views down the East River, but the Manhattan Bridge overhead still dominated the experience of the flat landscape. To give the ground level more presence several years later, MVVA raised the central lawn ten feet above grade, where it now draws together the views of the water, land, and the structure of the Manhattan Bridge. The hillside that rolls gently toward the shore softens the appearance of a colossal bridge stanchion a dozen or so yards out into the shallows of the East River. The landscape is now a welcome transition between the life of the street and the park interior, with a stair lined with bottlebrush buckeye to carry people up and away from the city into a calming zone by the river.

The most challenging landforms in Brooklyn Bridge Park, from both a technical and design perspective, were the three "sound berms." Unlike other areas sculpted for people to occupy, these extreme earthworks were added to reduce the noise in the park to acceptable levels by deflecting sound coming from the BQE. These three berms have become integral to the park experience, drastically lowering the noise while balancing the vastness of the harbor with something of comparable presence on the landward side. Blended into the park's interior, the hills create a prominent eastern edge that cradles the uplands and screens the view of the expressway.

Building landforms at this scale is costly. Typically in park construction, earth excavated from one part of a site is used to build a landform on another, a method known as balancing cut-fill. But the datum here was barely above the high tide line, meaning that all excavation, however shallow, would extend below the water table and invite water into the site. In any case, digging down would not have produced anywhere near the volume of soil needed for the monumental topography we were proposing. Meanwhile, purchasing all the required fill would have put a greater strain on the construction budget.

Fortunately, Adrian Benepe, then commissioner of the New York City Parks and Recreation Department, alerted us to a massive amount of pulverized rock being excavated by the Metropolitan Transit Authority for its extension of the Long Island Railroad from Queens to Manhattan's East Side. The so-called East Side Access operation, just four miles away, provided 75,000 cubic yards of clean bulk fill, enough volume to build the Pier 1 landform and a sizable portion of the sound berms. Because the city classified the tunnel rock as waste, we needed to cover only the cost of its transport, saving money for other park features.

PATHS: HORIZONTAL AND VERTICAL CURVES

MVVA considers paths the engine of a designed landscape. When laying out a circulation network, the design process starts by thinking about the alignment of walkways in plan, introducing curves that structure movement and enhance views. Manipulating bends in a path can make a tight space feel larger than it is and reveal destinations or landmarks in more compelling ways than results from a straight path. Of course, paths are also aligned to weave up and down, over crests and into valleys to make the most of the terrain. The interaction of the paths and the park's manufactured topography builds a rhythm of suspense and discovery as visitors move through the landscape.

The park's design minimizes the need for stairways while maximizing comfortably sloped paths to enable a much greater range of participation in the landscape. The Americans with Disabilities Act (ADA) limits walkways to a 5 percent incline before requiring handrails and landings. This is seen by some designers as a limitation but is in fact a wonderful generator of form, encouraging gentle slopes that are fun and comfortable at the same time.

The park's manufactured topography builds a rhythm of suspense and discovery as visitors move through the landscape.

A successful path network gives people options for moving through space: some routes are easier, some more challenging, and all engage unique aspects of the landscape. A stairway may offer a direct route between areas at dramatically different elevations, but a gently sloped path that meanders between the same two destinations offers a slower, very different experience along the way.

Path layouts that are conceived in three dimensions provide a welcome alternative to the regularity of the adjacent street grid. In section, they connect higher elevations to the datum of the piers while bringing parkgoers down to the water. Unifying this parkwide network of paths is the thirty-foot-wide Greenway, which connects the chain of discrete park spaces into a cohesive whole. It was critical that the central spine break away from the old datum of the docklands. A completely flat Greenway would have felt dull. Instead, its alignment is playful, rising and falling while edging toward some spaces and pulling away from others, all to make the most of the topographic change.

At one extreme, near the ruins of Pier 4, the elevation of the Greenway dips almost to the level of the constructed sandy beach. From this low point, where the waves lap just steps away, parkgoers headed toward the Brooklyn Bridge may choose a side path to skirt the stabilized water's edge past Pier 3, or head inland to climb to a high point along the sound berm, where the river is almost entirely out of view. The paths reconnect at the next low point: the edge of the Spiral Pool at Pier 2, which draws tidewater deep into the site and pushes the Greenway landward with a bold, energetic sweep. A few steps farther, the path splits again: a smaller walkway meanders into richly planted uplands that cascade down along the sound berm, while the main route arcs out toward the harbor, directing views toward the Brooklyn Bridge and then the Pier 1 salt marsh.

CARVING

As so much of the park's topography had to be added above the existing grade, it would be misleading to say that the many pockets of sunken and semi-enclosed spaces were "carved" into the site, even though that's how they appear. Simply adding promontories would not make a great park experience or provide relief from the overwhelming scale of its surroundings. People need opportunities to be enveloped in smaller spaces. To achieve this, we shaped the added earthworks to form plenty of intimate, contained spaces throughout the landscape.

In the Pier 5 uplands, the massive sound berm had enough space to bend rather than hug Furman Street for its whole length. On its north end, it pulls away from the street to leave room for the new park maintenance headquarters and wraps the edge of a sloped lawn. With a loose frame of shaggy evergreen deodar cedars, the inner lawn hovers at a comfortable distance from the lively Picnic Peninsula below and feels like a world apart.

Adding enough topography to enwrap similar spaces on the pile-supported piers would have required expensive re-engineering. But to expand the range of landscapes at the center of the park, we decided that Pier 3 needed to have the same leafy interiority as portions of Pier 1 or the uplands. To accomplish this, we built up the edges of a new bowl-shaped Central Lawn over a couple feet of stacked geofoam blocks, with a lower area in the middle of the pier. The foam is considerably lighter than earth, allowing us to place deeper soil around the edges of the bowl to accommodate a continuous border of trees. Around the perimeter of the pier, the concrete deck forms a continuous walkway around the skeletal frame of the deconstructed shed, a precise architectural line against the waves of plants at the periphery. The restored blue steel frames remain anchored into their concrete foundation walls, which create a playful relationship between the original elevation of the pier and the new landscape.

Every bit of reshaped land creates opportunities for the landscape. The prow of the Harbor View Lawn at Pier 1 provides perhaps the most direct embrace of the site's essential relationship to the water—to the monumental bridges, and to the city beyond. Yet every outward-facing landform has an interior side, too. Between the big lawns of Pier 1 lies the Vale, a beloved hollow set within an intimate frame of oaks, dawn redwood, and bald cypress that opens to the river on the west side. In some ways, the park is defined by its surroundings, but we came to understand that the landscape must also offer spaces where visitors are able to appreciate exactly where they are.

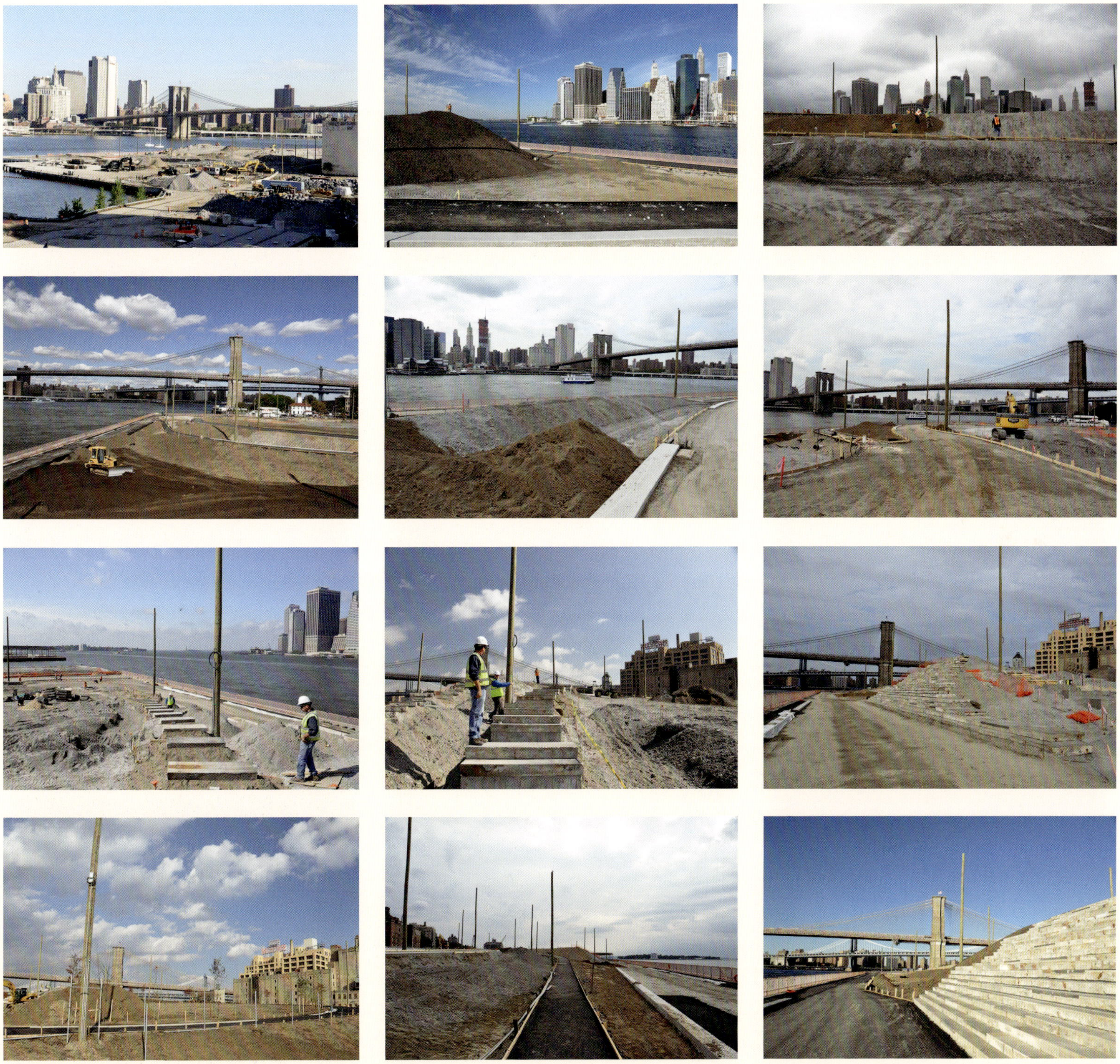

Pier 1 topography under construction, 2009.

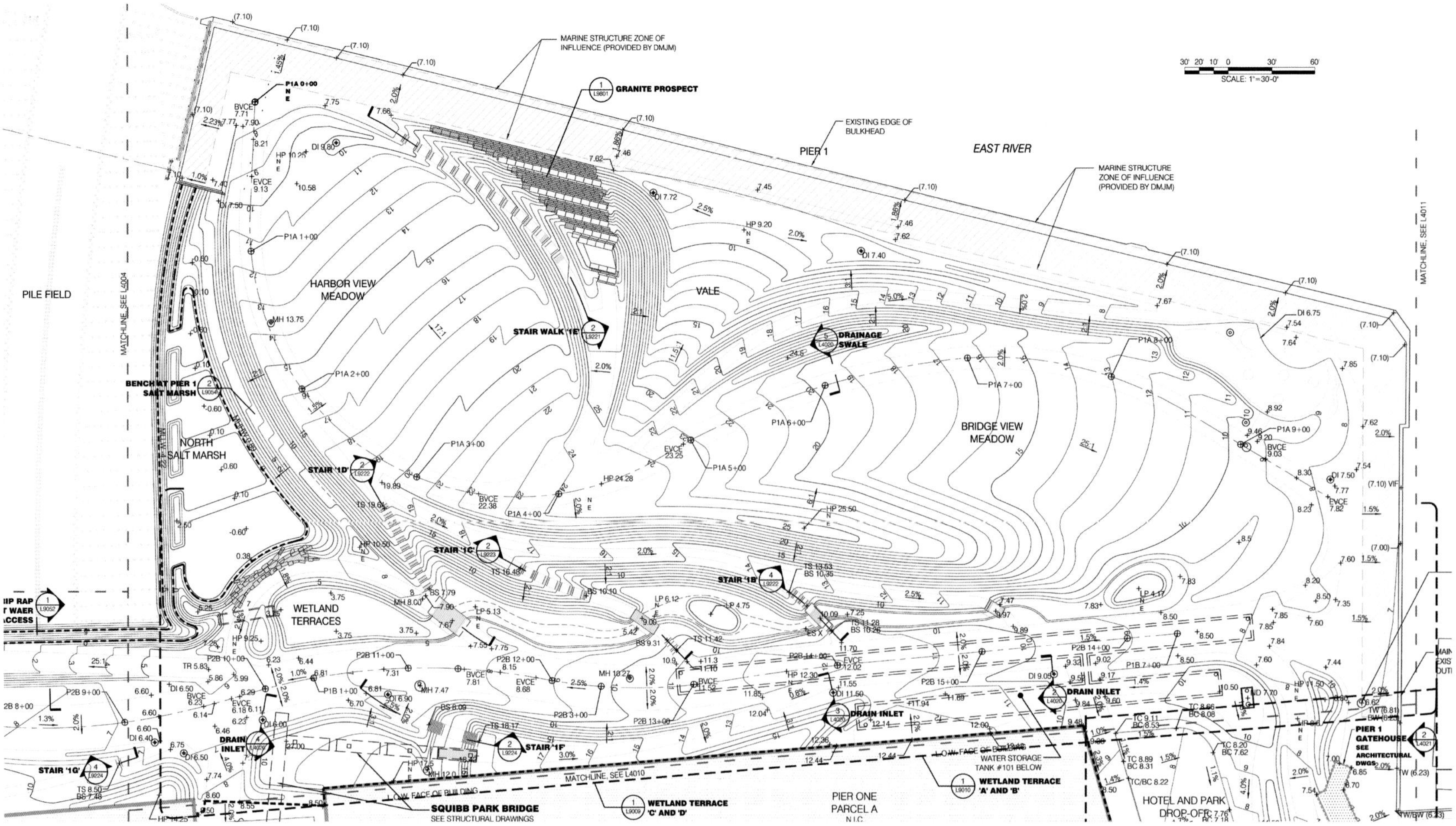

BREAKING THE DATUM

Added topography created a dynamic landscape that disrupts the existing elevation of the industrial waterfront.

LEFT *Pier 1 grading plan, 2008.*

RIGHT *An aerial view of Pier 1 after completion, 2010.*

Budget
Big screen. Bigger Imagination.
SyFy

HIGH POINTS

Extensive earthworks, like the twenty-foot-tall drumlin at Pier 1, provide the necessary height to emphasize views and differentiate space.

LEFT *A film screening near the Granite Prospect at Pier 1.*

RIGHT AND FOLLOWING PAGES *The vantage point from the top of the Granite Prospect.*

LEFT *The Pebble Beach at Main Street Park below the Manhattan Bridge.*

RIGHT *A view from above of the ten-foot-tall hill at Main Street Park.*

PATHS: HORIZONTAL AND VERTICAL CURVES

Walkways are threaded through the topography to give parkgoers flexibility in navigating the landscape.

LEFT *The inland paths of Pier 1 beside the freshwater wetlands.*

RIGHT *The Vale at Pier 1.*

DOGS
AND
FURNITURE
ARE NOT
ALLOWED
ON THIS
LAWN

CARVING

Smaller, more intimate spaces are cradled by the surrounding topography to balance the immensity of the park's surroundings.

LEFT *A model made at the MVVA office to study the integration of the sound berms and landscape.*

RIGHT *The Pier 2 uplands and sound berm.*

THE SOUND BERMS

Cut into a natural escarpment below the edge of Brooklyn Heights, the two-level Brooklyn-Queens Expressway (BQE) unleashes the full force of its traffic noise toward Brooklyn Bridge Park. This was less concerning when the waterfront hosted only warehouses and shipping activity, but for the park to function as a public space, the deafening roar of passing vehicles had to be mitigated. In 2003, the acoustic design firm Cerami & Associates measured sound on the site at eighty-five decibels—a level that makes normal conversation from more than a few feet apart impossible and causes hearing loss with prolonged exposure.

Cerami worked closely with MVVA on a solution for sound attenuation, first considering hard barriers much like the tall concrete walls often used to insulate neighborhoods from highway traffic. These ideas were rejected because they seemed unattractive and posed the risk of reflecting noise back into the neighborhood above, or even trapping vehicle exhaust on Furman Street below the highway. We briefly explored using low-rise buildings along the street as noise barriers, but lining almost half a mile of the site with even narrow structures would have used up precious park acreage. The team then considered "bio-walls"—stacked concrete containers filled with soil and lush planting—but the community saw these as an apologetic cross between a wall and a landscape. This led us to explore the use of earthen berms shaped at an angle to deflect noise up and over the park. These unusual landforms held the potential to solve our noise problem, screen the view of the highway, and add to the character of the park.

Cerami determined the shape of the berms to cast just the right "acoustic shadow" on the site. Sound waves, which expand and weaken as they move through space, reach park visitors at ground level far out on the piers at a tolerable sixty-eight decibels. If the angle of the berm was too steep, noise would bounce back in the direction of Brooklyn Heights; too shallow and the berms would fail to protect the park. Cerami determined that a berm would have to be huge, from twenty-five to twenty-seven feet tall, with its side toward the BQE angled at about forty-five degrees.

With those parameters in hand, we built physical models to determine how the sizable infrastructure could be integrated as elements of the landscape. A single berm along the entire stretch of the BQE would have been most effective, but due to concerns about the structural capacity of subway tunnels below, emergency access, and the need for entrances into the park from Furman Street, we decided to break it into three segments. The longest and first piece to be constructed begins just south of Pier 2, where a pair of subway lines run beneath the site, and continues on to an existing subway vent structure in the uplands of Pier 4. Across from Montague Street, the second berm to be built forms the eastern edge of the Pier 5 uplands and extends south to a park administration building and the entrance at Joralemon Street. The final berm defines the uplands of Pier 2 and heads north to another entrance from Furman Street, just passing the point where the BQE veers inland. Phased construction of each berm—earthworks like no other we had ever built—allowed us to improve installation and planting techniques as work progressed.

Typically, soil mounded to a forty-five-degree slope would slump shortly after installation, and the New York State Department of Environmental Conservation mandated that surface erosion be prevented to avoid discharging soil into the surrounding environment. Following the direction of the geotechnical engineers, we accepted a well-draining bulk fill provided by the Metropolitan Transit Authority to form the underlying shape of the berm, which would then be covered by twenty-four inches of planting soil. The bulk fill and the horticultural soil were held together between layers of geo-grid tiebacks, the kind of polymeric lattice used to strengthen retaining walls. As the new plantings established themselves, their roots would further increase the stability of the slopes.

Kelco Construction devised ingenious construction methods to meet the unique challenges of the berms. For the first segment, the steepness of the fill made it almost impossible for builders to access the slopes to install the top layer of planting soil. To overcome this challenge on the subsequent berms, planting soil was installed concurrently with the bulk fill. They called this the "lasagna method," creating room for their workers to stand on the highest completed level and then move upward incrementally as the berm rose. This allowed for continuous foot access during construction and eliminated the need for expensive motorized lifts. It also enabled us to seed a temporary cover of

What started as a strategy to mitigate noise has become a defining feature of the landscape.

plants to help control weeds.

Plant establishment and long-term erosion control on the steep slopes became ongoing challenges, but the peculiarity of the berms has made them laboratories for horticultural innovation. The period between seeding and root growth is particularly delicate: without protective measures, the combination of unnaturally steep grading and bare soil all but guarantees washout during heavy rain. We covered the surface of the first berm with jute mesh, reasoning that the loose weave would allow the seedlings to grow through gaps in the fabric while also holding the soil in place, but the thick jute blocked sunlight and impeded the tiny emerging plants. For the later berms, we secured the horticultural soil with geo-fibers, tiny, netted strands of polypropylene that mimic the stabilizing properties of roots while giving the plants time to grow.

To prevent seed from washing down the slope, we adapted a technique from regenerative agriculture that involved hammering dimples into the top layer of horticultural soil to make a matrix of small, shallow depressions in the hillside. Originally developed by farmers who observed the benefit of hoofprints in establishing grass on sloped pastures, this method held the seeds in place long enough for them to germinate and collected water to nourish the emerging plants.

The composition of plants that covered each berm was updated over the course of construction, too. After tracking the limited progress of warm-season grasses and native forbs on the first berm, we deployed a more durable cool-season mix of fescues, bluegrass, and autumn bent grass moving forward. Getting these species established was more easily coordinated with the berm's construction schedule because their germination occurs over a longer period. Vibrant green in the milder months and awash in earthy tones in late summer, the grasses offer a welcome contrast to the rest of the park's trees and shrubs. The second, southernmost berm was also planted with taller growing switchgrass at its base that suggests the scruffy texture of coastal dunes.

With years of testing and the added benefit of a close working relationship with the park staff, we were able to optimize maintenance practices. We cut a single diagonal path into each side of the first berm, but they provided insufficient coverage for maintenance and felt unsafe. The other berms were graded with several narrow, horizontal "goat paths" cut into the sides, making it easier for workers to gain footing while planting, weeding, and watering, all without compromising the berm's sculptural form.

Unexpectedly, the berms created habitats for many types of wildlife, including some that rarely thrive in an urban environment such as Brooklyn. The linear grasslands of the berms, edged at their base with thickets of trees and shrubs, form an almost seamless corridor for animal movement across the eastern half of the park. Nesting and foraging animals have ample cover and protection from human disturbance, as the slopes are too steep for regular foot traffic. Park staff found the nesting grounds of the rare yellow bumblebee at the base of the berms and must check the land for nesting mallards before cutting back grasses. The abundant grasses also host skipper butterflies that flit about the berms all summer. For New Yorkers who pay attention to their non-human neighbors—and there are many of them visiting the park—the bustling activity of life on the berms is evidence that the landforms have earned their place on the Brooklyn waterfront.

What started as a technical strategy to mitigate noise has become a defining feature of the landscape of the park. Over time, we realized that the berms reinstated an image of the bluff that connected Brooklyn Heights to the shoreline—a natural feature distorted over centuries and all but erased by the BQE. The sense that the berms somehow belong to the larger landscape comes from their being an entirely practical design response to make the park habitable. Even their eccentric curves resulted from a particular need to dodge incompatible site conditions or accommodate park infrastructure. They lend their bold scale to the park interior, creating cool microclimates and quiet backdrops for every kind of activity while offering momentary but profound separation from the city.

The berms are huge, even imposing, but they are a fitting response to the scale of the expressway—an answer to the problematic legacy of hostile highway infrastructure in a new era of more humane city building.

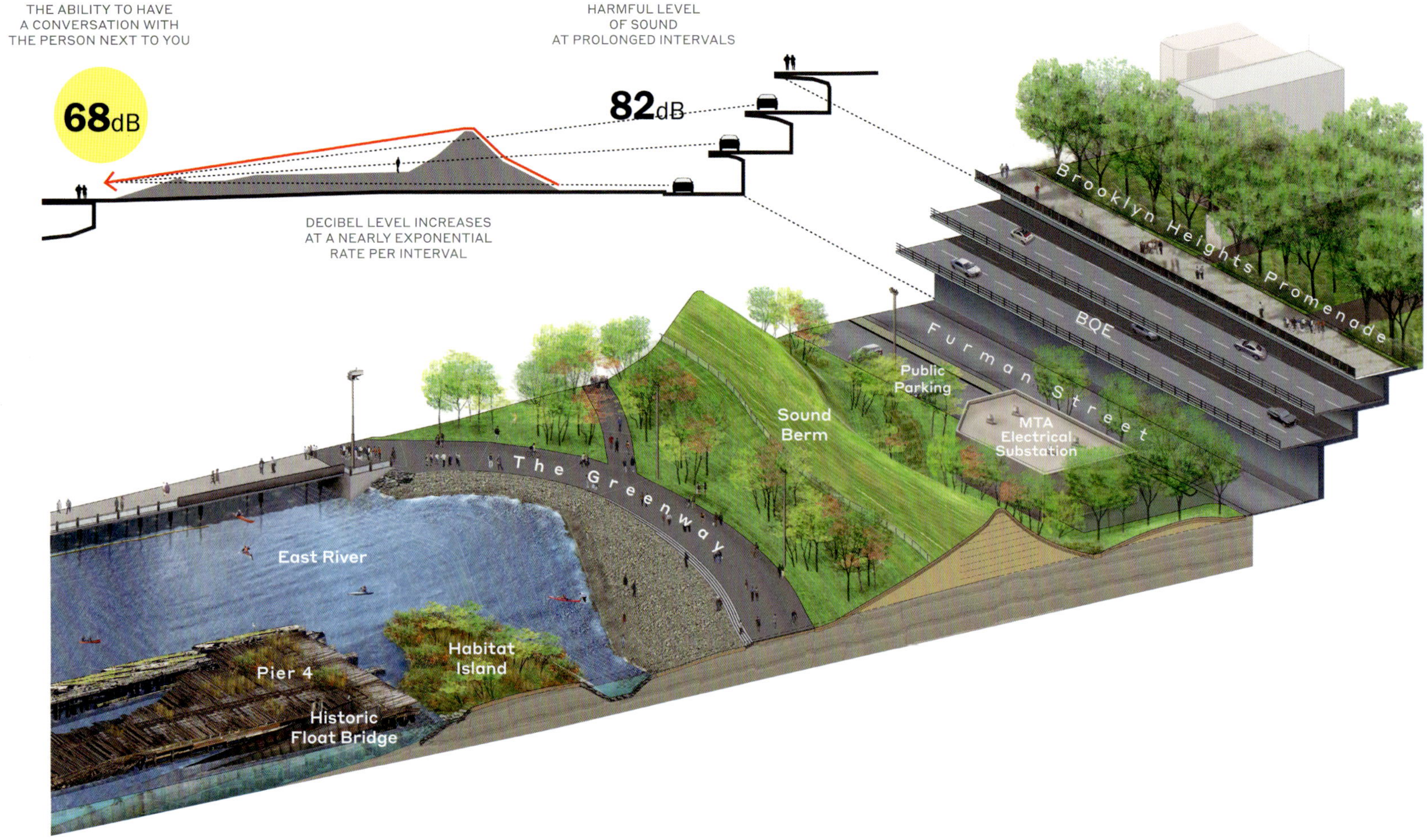

ENGINEERING THE SOUND BERMS

To tame the noise from the Brooklyn-Queens Expressway, we built a series of twenty-five-foot-tall, noise-abating landforms.

LEFT *An axonometric perspective and section of the sound berm near the Pier 4 beach, illustrating noise mitigation.*

RIGHT *View from the Brooklyn Heights Promenade of the sound berm at Pier 5 nearing completion, 2016, and an aerial of the completed Pier 5 uplands, 2021.*

Park Maintenance Center
Pier 5 Uplands
Furman Street
Maintenance Paths
SOUND BERM
BQE (Brooklyn-Queens Expressway)

PLANTING
THE
LANDSCAPE

PLANTING THE LANDSCAPE

When we first started scouting the future park site, it was an abandoned port, flat and largely paved. Here and there, weeds and volunteer saplings had forced their way through cracks in the concrete and asphalt, and a few trees stood tall. One was an Asian princess tree growing just above the high tide line adjacent to Pier 2; another, a towering North American cottonwood, spread its limbs not far from where Jane's Carousel stands today. Once completed, the park had 1,000 trees, among which those two survivors have maintained an honored place. In the early years of planning, they gave us hope. Seeing the princess tree's explosion of purple snapdragon-like flowers in May and the cottonwood's yellow glory in November helped us imagine how alive the park could become.

Our encounters with those volunteer trees revealed a guiding principle: that a parkgoer's experience of plants is personal. Planting must welcome park users, orient them, and guide them to different activities. Plants not only perform as forms structuring space but also as living beings with their own ways of expressing themselves. They are approachable, touchable, smellable, tasteable. Much more than scenery, the plants invite visitors to cultivate their own relationship with the landscape.

A convention in park planting suggests that heavily used areas should prioritize simplicity to make maintenance easier. We disagree, believing instead that places with concentrated activity need the benefits of complex plantings as much as sheltered, quiet places do. We worked to turn the urban shoreline into a stronghold of biodiversity with constructed habitats, and rather than separate these from human activities, we made sure they were accessible to enrich each visitor's experience. Where we planned for high human impact on lawns, terraces, and paths, the character of each space grew out of the diverse, layered plant communities at its edges. We sought a form of designed nature that welcomes parkgoers—ecosystems that include people.

To inspire us about how the landscape might feel, we pinned up Eric Drooker's cover of the August 5, 2002, *New Yorker*. His illustration depicts a commuter emerging from an underground subway station, wide-eyed, into a jungle of trees, ferns, vines, and shrubs. The scene brought us back to that unexpected first encounter with the volunteer cottonwood and princess trees, perfectly capturing the feeling of awe that comes from suddenly being face-to-face with nature in the city. By this time in the office, we were using the term "hypernature" to describe the strategic placement of dense plantings in landscapes that enable a potent experience of flora and fauna that is less common in urban settings. Today in the park, miniature forests, constructed salt marshes, a small wetland, wildflower meadows, and beaches lend vibrancy to every type of activity.

Hypernature creates wonder for parkgoers because interactions with this kind of planting feel spontaneous and come with a sense of discovery. From this dynamic grew a deeply informal design sensibility for the park. Of course, informality alone is not enough to generate a coherent set of aesthetic principles for a designed landscape. To unlock a sense of place specific to our site, we played the inherent wildness of plants against the geometric rigidity of the port infrastructure. Inspired by the disintegrating piers we saw in South Brooklyn, where the unruly, twisting stems of volunteer sumac thickets were defying the steel and concrete edges, our goal was to heighten the contrast between the manufactured and the natural in the new landscape. Some of the park's first renderings in 2002 depicted big stands of rangy sumac with their radiant fall color playing against the hard lines of the piers.

Our plant palette was a cosmopolitan mix of native and adapted species, reflecting the natural history of the region and the cultural history of New York City. Many of the plants had loose, idiosyncratic forms, and we composed them irregularly, anticipating that over time each species would fill different spatial and ecological niches. We sought to embrace the uncertainty of how the plants would change as they occupied the site and welcomed many surprises. If plants were experienced as co-authors of the landscape, rather than components of a geometric or legible pattern, we hoped park visitors might be more likely to enter into their own conversations with their environment.

PLANTS ORGANIZE THE PARK

The immense scale of the urban frame surrounding the site was a gift that came with challenges. From within the park, people needed to be able to feel the expansiveness of the city but also have the choice to feel the sense of interiority that comes from a closeness with plants. The public funding for the park was released in increments by the City and the State of New York throughout phased construction that spanned thirteen years. This required a high-level concept to organize the entire landscape, as well as strategies for the first plantings to stand on their own and merge with later phases. Flexibility was essential in order to respond to unexpected changes over the coming decades.

The planting design clarifies the spatial organization of the park as a whole. Given the site's long and narrow footprint, our designs had to be easy to understand and consistent across time and space. To define the primary circulation corridors, we invented "complex hedgerows"—long, thick bands of tall trees combined with understory trees and groundcovers to amplify the density of the path edges. For example, closely spaced mixes of river birch, yellowwoods, clump-form coffee trees, and pin oaks reach over a mid-level of redbuds and Green Shadow magnolia, with a cacophony of summersweet, shrub roses, switch-grass, and columbine that provide texture at the ground level. The effect of the complex hedgerow is like that of a compressed slice of a much larger landscape. Simple rows of trees may have been enough to trace these important park routes, but dense and textured borders, with plants that billow over fencing or spill out onto paths, make the slender waterfront feel endlessly deep.

For a hedgerow on Pier 1, we thought to plant multi-stem trees that could increase the visual density of tree trunks, but no plant grower in the region was selling what we wanted. We convinced Halka Nursery in New Jersey to plant saplings of one species in clumps of three or four, fusing their roots together to create one multi-stem tree. This concept transformed familiar single-trunk urban trees—planes, coffees, or honey locusts—into a new and exuberant form.

Counterpoints to hedgerows that direct movement, park lawns are places to pause. As the primary venues for open-ended activity, they needed to be of many different sizes and well distributed. We wanted all of them—whether flat, sloped, wide open, or tucked away—to feel connected to the vegetation around them. For instance, the lawns on Pier 6 are dotted with thickets of plants we saw thriving at other shore sites—junipers, Bosnian pines, sassafras, and tupelo, with bayberry and a few shrub roses at the sunny edges. On the leeward side of the thickets, rough-hewn granite blocks offer informal seating. Sheltered from harsh winds coming off the water, these shady areas are popular with caregivers watching their kids at play. In contrast, the two one-acre lawns on Pier 1 are sloped and wide open, providing large gathering spaces and long views. They are edged by thick continuous bands of oaks, multi-stem plane and coffee trees, with textured understories of hawthorns, sumac, and closer to the ground, sedges, woodland asters, fragrant viburnums, and wild geraniums. A few sweetgums and low-branching American hollies here and there in the wooded borders make the lawns feel like they are carved out of a rich habitat.

The overall planting strategy balances organizational clarity with a reading of the park as a found landscape. We avoided creating hard lines and steered the design away from recognizable patterns and geometries. The only large-scale rectilinear forms in the park are the preexisting piers, bulkheads, and warehouse frames. Every plant edge curves and meanders, and the plants themselves are selected and maintained to grow in the same loose forms as they would on their own. The advantages of this approach are well known. We had in mind, for instance, how Prospect Park's undulating Long Meadow bends out of sight behind naturalistic clusters of trees, making it feel limitless by hiding its full extent from view.

The curves and informal groupings of plants have also enabled seamless transitions between construction phases, whereas more ordered compositions would have made these edges more conspicuous. By not limiting the design to prescribed styles or to a single plant palette park-wide, we were able to explore new ideas as the park evolved. The adaptations were able to respond to the harsh natural forces of the water's edge and the constantly changing needs of a large and diverse park user community.

While the informal planting design echoes natural systems, it also absorbs some of the unexpected results that come with introducing one thousand trees and tens of thousands of smaller plants into an urban waterfront. The native pitch pine, for example, seemingly perfect for these site conditions, mostly failed. On the other hand, the stag-horn sumacs on the wooded hill along the wetlands of Pier 1 grew twice as tall and twice as wide as we had imagined. Far from stunned, we love that they reach out over the ponds, further blurring the edge between the forest and the small channel in ways we could never have planned.

STRUCTURE AND SENSATION

A "planted form" is an arrangement of plants that follows a recognizable pattern to serve a particular purpose, much like architects use typology to draw connections between scale, form, and programmatic function for similar building types. In the park, planted forms include groves that shelter, allées that orient and direct, waist-high flower meadows that dazzle and occupy space without blocking views, and trimmed hedges that define outdoor rooms.

While the organizational structure of a park needs internal consistency to maintain its coherence over time, planted forms are expected to evolve as plants grow. The sheltering space enclosed under the low branches of young catalpas with their singular calligraphy is dark and intimate in summer. But after their big, heart-shaped leaves have fallen, the gnarly branches let the sun in. Over decades, the quality of the space continues to change as the canopies stretch higher.

Planted form engages the space-making capacity of plants while also accounting for the ways they change. We consider the ephemeral seasonal qualities of plant species as intertwined with the felt qualities of planted forms. If "planted form" refers to the spatial composition of plants, a "plant palette" is the total combination of visual, auditory, olfactory, and tactile sensations produced by one range of species over time. Plant palettes are ensembles of species, grouped together to become more than their individual expression. Ginkgo and red maple create a very different impression together through the year than ginkgo mixed with the evergreen Japanese cedar.

We think of the spatial distribution of planted forms and their plant palettes as snapshots of structure, color, and texture composed to engage the imagination. As we work on planting design in the studio, we shuffle photos of plants with either cut paper or digital assemblages to evaluate combinations of leaf, bark, branch, and flower across the seasons. Even the most experienced plant lover cannot imagine all the ways the mingling of plants will play out over time. Unlike modeling software, this hands-on method of making collages gives us limitless possibilities for testing ideas without the false impression that we know exactly what to expect. The design process remains open, and much like the natural forces at play in the built landscape, chance plays a large role in how the design evolves.

Intuition continues to drive the design process even after the chosen plants have arrived on-site for installation. During the final stages of construction, we stand by and reserve the right to revise final placement. Layout on site provides the design team with one last opportunity to see in real time the relationships among species when they are mixed, with the unfolding personality of every single tree, shrub, and perennial on full display.

EXTREME MEASURES

Throughout planting design, we had to plan for amended growing conditions to achieve many of our desired results. No part of the site could have supported the variety of habitats that thrive there today without some major technological intervention. These included massive earthworks, constructing a new shoreline, formulating seven different types of horticultural soils, and implementing drainage and irrigation systems—all with the support of expert technical and scientific consultants. And today, the park's complex ecosystems are supported by a year-round staff of nineteen full-time gardeners and dozens more in the summer.

Enabling large trees and meadows to thrive on the concrete decks of Piers 3 and 6—landscapes that now feel as inevitable as old pastures—was a triumph of horticultural technology. Sitting on piles above the water, the concrete deck of Pier 6 could not have supported the weight of soil that was installed for the plantings on Pier 1, but we were determined to create something equally immersive. We initially considered a perched wetland, which later evolved into the more simply engineered Flower Field. The drifts of native and adapted flowers with lower water demand have thrived—some now grow eight feet tall every summer—in the shallow soil, exploding with seasonal color and buzzing with pollinators. By pushing what was technically possible on top of a concrete deck suspended over the East River,

We played the inherent wildness of plants against the geometric rigidity of the port infrastructure.

this half acre of goldenrod, cardinal flower, rose mallow, aster, rudbeckia, and more than forty other species feels surprising and authentic at the same time.

ADAPTATION AND CHANGE

Planning for adaptation is at the core of planting well. It extends beyond design and construction into long-term management. Most of the plants in the park were installed quite small, giving them time to acclimate to site-specific conditions as they grew. Specifying younger, less expensive plants freed up funds to develop high-quality manufactured soils and support for other measures that ensure longevity. We followed the gardener's adage of "plant a one-dollar tree in a five-dollar hole." It was energizing to know that those plant communities that we were setting in motion, and the park environment that sustains them, would likely face both predictable and unpredictable changes as they grew and delighted many for years to come.

It was inevitable that this site, like the rest of coastal New York City, would face potentially catastrophic flooding made worse by the climate crisis. Most of the site is only a few feet above sea level, so the park was designed to let floodwaters in rather than attempt to keep them out. We calibrated the new soils accordingly and selected plants that tolerate saltwater inundation, many of which evolved in similar coastal environments. If certain shrubs, grasses, and perennials—bayberry, switchgrass, or goldenrod, for instance—call to mind the rustic beauty of a back dune on a nearby New Jersey beach, it would stand to reason that they would also be equally capable of living with salt spray, flooding, and the sandy, porous soils that were installed at Brooklyn Bridge Park.

The park's trees, too, were selected for their records of success during saltwater floods and then installed above the 100-year floodplain. But after Hurricane Sandy topped that elevation in 2012, we discovered which tree species lived up to their hardy reputation: honey locusts, lindens, southern magnolia, and bald cypress did very well, aided by park staff who flushed their roots with freshwater repeatedly. London plane trees and dawn redwoods unexpectedly recovered less reliably, providing a practical lesson that we carried forward into subsequent stages of planting design. We also anticipated how the hedgerows—densely planted at the start for immediate effect—would require incremental thinning over time to avoid unhealthy crowding and support natural succession. A variety of oak species is dispersed across the park. Slower growing at first, the oaks will eventually provide an enduring structure for the landscape. Meanwhile, the layered understories were initially planted with more sun-loving species, which were shaded out surprisingly fast and replanted with plants that thrive with minimal sunlight.

Brooklyn Bridge Park staff have stepped in to guide many of these changes. They practice a kind of ecological horticulture that fosters the initial establishment of a plant community and encourages development in a general direction but otherwise allows species to interact and fill distinct roles as they naturally do. This requires constant monitoring and analysis, activities that are made visible to the public, which helps build deeper bonds between the park and its visitors. On Pier 1, the first to be planted, sweeps of shade-loving mayapple and bluebells have been tucked in by the gardeners along with a small and graceful striped maple, a tree that generally does not thrive in New York's hot summers but so far loves its place in the understory. Rather than remove a few volunteer red maple seedlings that had emerged, the gardeners began to nurture them. The constant play of natural phenomena makes the park experience casual and different every single day.

A RELATIONSHIP WITH PLANTS

Change is not just the practical reality of plants but also a key part of their wonder. We knew that experiencing plants would become a way for parkgoers to attune to changes in the natural environment, which in turn connects them with a world beyond the daily routines of city life. For many people, the park would be their primary experience of a living landscape. Knowing this, we wanted the growth as well as the day-to-day seasonal shifts of plantings to be as evident as they could possibly be. Twice each day, tides roll in and swallow up the smooth cordgrass in the salt marshes; in early winter, fragrant viburnums change from green to purple and then drop; and in spring, horse chestnut leaves are the first gauze of bright green on the waterfront. To

make the old concrete and asphalt port truly come alive, we needed carefully chosen varieties of plants, but we also need to cultivate whole ecosystems, with all the willful untidiness they bring.

By definition, a park is not nature. The magic of a park is that it enables a personal relationship with nature that is not available in nature. The serendipitous moments when people can commune with their environment happen because Brooklyn Bridge Park has become a part of their everyday world—it is green and living, but still connected to the life force of the city. The most meaningful experiences of the park often occur when the comforts and joys of daily living relax our imagination enough to receive them: pausing on a jog to take shelter in the cool, shaded wetlands; cutting through a Juneberry thicket for a snack on the way to a summer pickup game of volleyball; gazing at the Manhattan skyline across the copper-colored salt marsh cordgrass; or glimpsing the lights of Jane's Carousel through a towering curtain of yellow cottonwood leaves.

HYPERNATURE

We were inspired by Eric Drooker's August 5, 2002, cover of the *New Yorker* titled "Last Stop," in which a man emerges wide-eyed from the subway into a jungle.

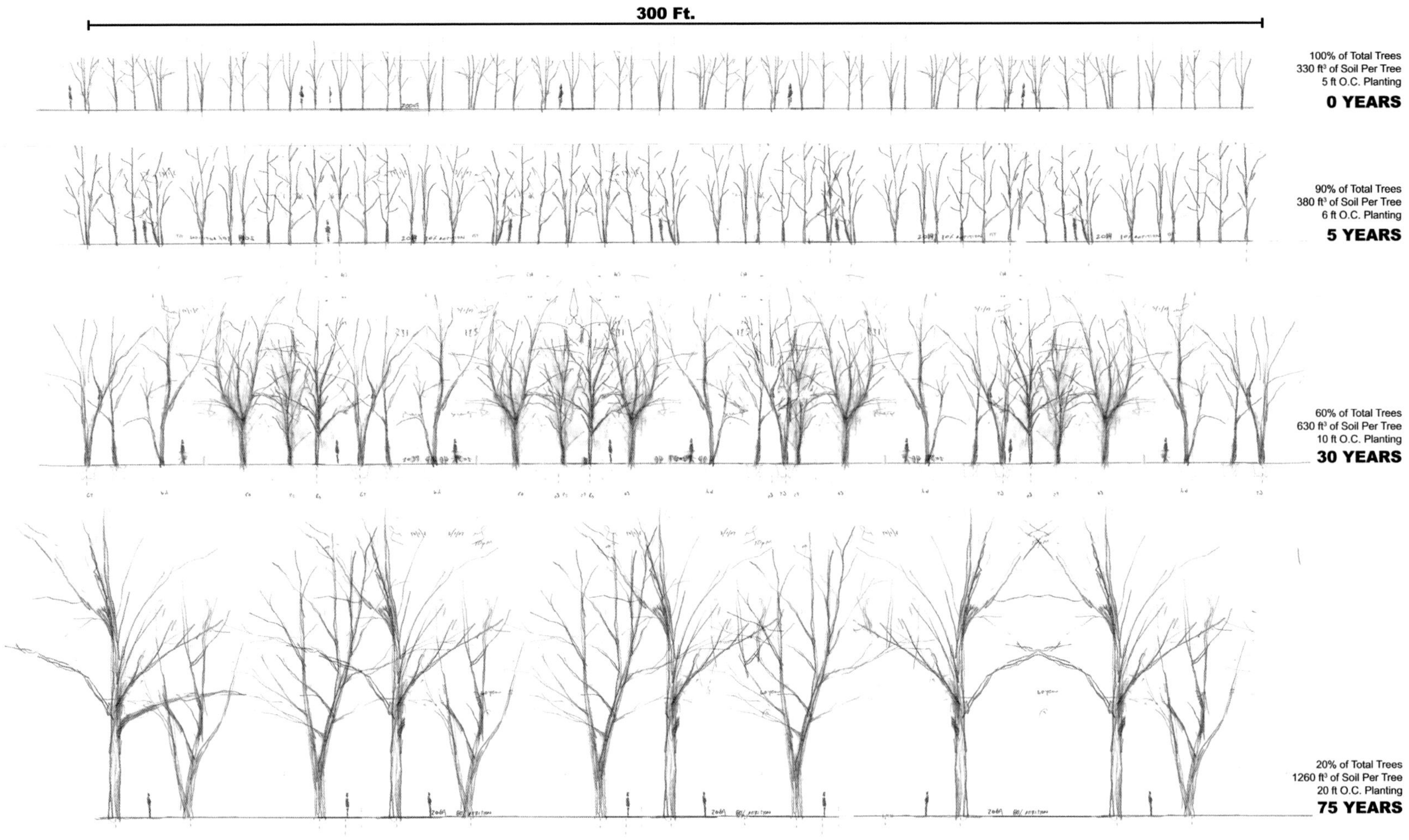

MANAGED SUCCESSION

MVVA projected the height and distribution of the trees that compose the park's complex hedgerows to anticipate how the landscape would evolve over the next seventy-five years.

PREVIOUS PAGES *An upland walking path separated from the Greenway by a densely wooded border.*

RIGHT *The Bridge View Lawn at Pier 1.*

Hedgerows separate narrow side paths from the Greenway, left, and shape sight-lines throughout the uplands, right.

Inventive soil engineering allowed MVVA to create thriving ecosystems across the park. At the Pier 6 Flower Field, half an acre of more than thirty species of flowers and grasses has grown from a thin layer of soil.

PAGES 160-165 *The Pier 6 Flower Field.*

PIER THREE

Completed in 2018, Pier 3 was the last of the major pier landscapes to be constructed. With the high-activity Pier 2 courts to the north and the Pier 5 soccer fields to the south, this pier provides less prescribed and more passive uses as well as the enveloping experience of rich plantings. On Pier 3, parkgoers can pause, relax, and enjoy the sensory pleasures of plants, water, air, and sunlight. The landscape is a sibling to those of Piers 1 and 6, where hedgerows and thickets define lawns for people to gather and take in the park's dramatic views.

But Pier 3's Central Lawn offers something different. While other parts of the park often draw the eye outward to landmarks, watery expanses, and distant shores, this one pulls one's focus inward to the landscape and companions close at hand. To accomplish this, paths and programmed spaces were kept to the periphery, surrounding a large open lawn, graded as a shallow bowl that rolls up gently at its meandering edges. Along its margins, which have a deeper soil profile, we planted clusters of trees, shrubs, grasses, and perennials to compose an ecotone—or a transition from one ecosystem to another—in this case the gradient of a grazed pasture to a riparian woodland. We chose plants that tolerate salt spray and wind: Chinese juniper, white cedars, and American hollies; coffee trees, sassafras and oaks; and many coastal meadow grasses that trace a soft, meandering fringe around the scattering of trees at the edges. Walking the wide concrete perimeter positions visitors between open water and a wild wall of green poking through the preserved steel frames of the pier shed.

The pier's extreme exposure to high winds and sun determined our plant choices and compositions. Near the water, plantings were designed to grow shorter and fuller, with a dense grove of rugged junipers that hunker down against the wind at the pier's harbor end. Tough coastal species like salt shrub, bayberry, and native roses billow out between the junipers. Nicknamed "The Cliffs," the garden's sinuous concrete walls form a series of niches that shelter visitors from the chilling harbor breeze and retain waist-high plant beds. These rise a few feet above the original pier deck, immersing visitors in a labyrinth of thick foliage. The tiny, bright pink flowers of sea thrift and cheddar pinks bloom in early summer, followed by late-flowering mounds of Montauk daisies. Sea kale flops its silvery, succulent

While other parts of the park draw the eye outward to vast planes of water and iconic landmarks, this pier pulls one's focus inward.

leaves over the walls' raw concrete sides. The beds produce clouds of white flowers that turn into pea-sized seeds held in finely branched stalks. These coastal specimens cannot be found anywhere else in the park, which helps emphasize the specific environmental conditions on the southwest corner of the pier.

A wide range of seasonal expression makes the pier landscape inviting throughout the year. Parkgoers may be pleasantly surprised with yellow vernal witch hazel flowers in January, and by March they may notice the blooms of Cornelian cherry dogwood as spring bulbs push through the leaf litter. When the Central Lawn is awash in the green of summer, the garden perennials at the Cliffs—a mix of sea holly, wine-red yarrow, and creeping geraniums—are aglow in the brilliant sun. Pier 3 is at its most colorful in the fall, when the orange and red hues of sassafras and chokeberry and the brilliant yellow leaves of linden are on full display. The coming of winter returns the park to subtler hues, as the glossy green of evergreen hollies shines against leathery brown oak leaves that stubbornly hang on into the new year.

On the northeast side of the pier is the Labyrinth, an exploratory play garden with a grove of quaking aspen that moves around enigmatic play elements that incorporate mechanics of light and sound. As the wind passes through the aspen canopies, the dapple of light and the sound of fluttering leaves add to the sensory landscape. Fragrant trimmed hedges of summersweet and mock orange shrubs curve along the paths, creating a maze of foliage just tall enough to feel both safe and exciting for children.

Along with the richness and variety in these thick margins on Pier 3, it was important not to complicate the simple narrative of transition from field to forest. Looking out from the Central Lawn, the aspens are a fitting backdrop to the veil of oaks and coffee trees, which conceals the whimsical play equipment. Many of the species are familiar components of landscapes across the northeast, especially those beloved and untidy farm fields tucked into forests. Every experience on the pier flows from seemingly chance relationships among plants and topography, each leveraged to improve visitors' comfort. Several lone oaks slip out from the wooded edges to cast shade on the boldly undulating lawn, which creates cozy pockets for gathering that shelter visitors from sun and wind.

When we began planning the park, we felt pressure to design most of the piers in service to the views across the water. But after opening Piers 1 and 6, we learned that visitors cherish the sublime beauty of glimpsing Brooklyn Heights and Lower Manhattan through a veil of leaves, too. As the trees at the outer rim of the Central Lawn continue to mature, the planted windows to the city beyond will become increasingly beguiling, while the soft lawn bowl—ever more contained by green—nurtures many parkgoers' desire for a pastoral escape from worldly cares.

Meandering plant beds surround the Central Lawn on Pier 3 to create intimate gathering spaces.

THE FRESHWATER WETLANDS

Transforming eighty-five acres of impervious pavement into verdant landscape was a rare chance to benefit New York City's stormwater infrastructure. More than half of the city relies on systems that convey both rainwater and sewage together in a single pipe for wastewater treatment, which overflow during heavy storms and discharge polluted runoff as well as untreated waste into waterways like the East River. We recognized that the park not only needed plants that could slow down and retain more water on-site but also a high-capacity system to help reduce overflow events. Throughout planting design, we also sought more efficient methods of care, since establishing the miniature forests, hedgerows, lawns, and meadows throughout the park would impose new pressures on the city's water supply.

On site visits, we discovered a decaying bulkhead in front of the National Cold Storage warehouse that was constructed long before the underside of Pier 1 was filled. This weakened gap at the junction of the pier and uplands provided an opportunity to excavate below the datum and create a lush, low-lying wetland ecosystem that addressed both stormwater and irrigation concerns. A narrow chain of ponds on the east side of Pier 1, the freshwater wetlands were designed to support the sustainability of the new landscape as much as the health of the East River. Today, they make up one of the most alluring and biodiverse spaces in the park.

We knew the one-acre Bridge View Lawn on Pier 1 would shed a significant amount of runoff. Although lawns retain some water, the soil beneath would be compacted by foot traffic, limiting absorption. With this in mind, the first wetland pond was located at the base of the slope on the pier's northeast corner, where it would easily gather runoff. Organic maintenance strategies developed by the park's staff reduce contaminants in the runoff, ensuring that the collected water is suitable for irrigation.

The wetlands consist of a 650-foot-long series of gravity-fed aquatic cells, including three ponds, threaded together by a densely planted rivulet. These are carved into the ground just below the elevation of the Greenway to direct water south, and each step in the chain contains a distinct plant community. Emergent pickerelweed and water lilies poke through the center of the first, uppermost pool, which also has sun-loving shrub willow, tupelo, blue

The wetlands offer an especially wide band of biodiversity and engrossing fullness within a slender stretch of the park.

flag iris, sensitive fern, swamp milkweed, Joe Pye weed, and cardinal flower at its edges. When water overtops that pond, it flows to a second, shaded pool bordered by coarse sedges, swamp rose, and ostrich fern. The third and lowest pool has little visible standing water. It brims with a thick patchwork of bulrush, cattail, and rose mallow, which acts as the final filtering element before the water drains into a 100,000-gallon underground tank. It also provides a visual transition from the diverse flora of the freshwater wetland to the monoculture of smooth cordgrass in the salt marsh just a few steps away.

Four entrances into Pier 1 cross the wetlands, each giving visitors a chance to access a freshwater ecosystem that is rare in Brooklyn. Despite their narrow size—just fifteen feet wide at one point—the ponds and plantings are home to an astonishing variety of creatures. The area is a crucial nesting habitat for wetland-dependent birds like the red-winged blackbird, a breeding ground for such insects as dragonflies and specialized hibiscus bees, and a resting and drinking source for migratory birds, swallows, and bats. On evenings in July, the nearby lawns flicker with fireflies, which depend on the moist, shady underbrush for most of their lifecycle. At other times, visitors might get to watch basking turtles, see herons stalking fish in the shallows, and hear the bellow of bullfrogs. The park provides plenty of unexpected encounters with nature in the city, but the wetlands offer an especially wide band of biodiversity and engrossing fullness within a slender stretch of the park.

The natural exuberance of the area is no accident; its construction required rigorous technical coordination to sustain a specialized ecosystem. We collaborated with eDesign Dynamics, developing tailor-made soil mixes that support the plants while optimizing water retention and filtration. Two layers of manufactured soils were placed above a sand drainage layer that rests on a liner of geo-synthetic clay. Stormwater percolates through the soil as it nourishes the roots, mimicking the filtration that would happen in a natural environment, until it is stopped by the clay liner and flows to the storage tanks for irrigation.

The dramatic topography of Pier 1 provides a tall tree-covered hillside that nestles up against the wetlands, amplifying their depth. Flecks of sunlight within the cool shadows, the sounds of insects and birds, and the aromas of foliage and flowers produce a moment of distance from the city. Visitors slip past a narrow gap in the enveloping juneberry trees along the Greenway to cross one of the steel-grate footbridges that pass over the rivulet, pulled deeper into the Pier 1 landscape by the quiet of the narrow transition. Hidden in plain sight, the green infrastructure suggests a new direction for city parks where functional ecosystems feel present, in all their woolly wildness.

RIGHT *A pool within the freshwater wetlands at Pier 1.*

FOLLOWING PAGES *A section sketch of the freshwater wetlands by MVVA.*

charrette
16.25
CF
RM
MV
IV
THIN + WHIPPY THICKET
Betula
Cotinus obovatus smoketree 30'
Acer pensylvanicum
Cercis canadensis
Ostrya virginiana 30 - 50' older corridor
Corylus cornuta shrub, leggy 15'
Hamamelis virginiana
Ribes hirtellum golden currant leggy shrub 15'
Sambucus racemosa shrub tree
Staphylea trifolia flower shrub/tree
Amelanchier grandifolia
MOIST LOW THICKET
MV Magnolia virginiana
RM Rhododendron maximum
HT Halesia tetraptera 15'
CF Calycanthus floridus 7'
FM Fothergilla major
VA Viburnum alnifolium
IV Itea virginiana
charrette

1/4" = 1'-0"
WETLAND 'C'

PREVIOUS PAGES *An elevated walkway through the wetlands on Pier 1.*

LEFT *The wetlands leading out to the salt marsh at Pier 1.*

REMAKING THE WATER'S EDGE

REMAKING THE WATER'S EDGE

By the 1950s, the area of the East River just south of the Brooklyn Bridge had entirely lost the gentle slope of an ecologically rich intertidal zone, replaced with the hard edges of concrete seawalls. There were virtually no opportunities to appreciate the tide's ebb and flow, to experience wildlife habitat, or to dip a toe in the water. The shoreline north of the bridge was only marginally better, with a few makeshift paths down to the shoreline. We set out to change that, approaching the need to replace the abandoned shipping infrastructure as an opportunity to bring the East River into the life of the city. There were some signs of progress. Water quality in the East River had improved drastically, and we often saw kayakers out on the river during visits to the piers. We were determined to build a resilient edge that would be good for the river and even better for the people who use the park.

For decades, much of Brooklyn had little or no direct experience of the site or its relationship to the East River. It could almost exclusively be seen from high above on the Brooklyn Bridge, the expressway, or the Brooklyn Heights Promenade. Even so, in community meetings, a common thread emerged: people wanted to hear the water lapping against the shore, smell its salt, and perhaps even feel the water wash across their skin. In our first public meeting, the appeal from longtime resident Mickey Murphy stuck with us. An octogenarian, her most keen desire was to be able to step into the East River. More than a wake-up call, it was a sign of the changing expectations for public parks in New York City. Mickey helped us realize we could turn the water into more than just scenery by rebuilding the waterfront and encouraging experiences of the river unknown to most living New Yorkers.

Nevertheless, the water's edge could not be remade wholesale. Too many layers of marine infrastructure were embedded in the site, and the budget was limited. We also wanted to preserve the maritime legacy of the waterfront and the existing piers—nearly a third of the park's total area—which presented a rare opportunity to bring activity to the water. Bulkheads and platforms in good condition were retained to maximize the usable area of the park, while degraded sections were transformed to emphasize the presence of the tides. Our commitment to recovering the water's edge for people, combined with an economical approach to reusing existing infrastructure, led us to three main types of shoreline conditions: industrial edges that could be preserved and activated with new program; new stabilized banks of stone riprap known as revetments; and dynamic tidal zones featuring constructed marshes, boat ramps, and beaches.

PRESERVED AND REPURPOSED

The marine structures that composed Piers 1, 2, 3, 5, and 6 were all preserved. Except for a few acres of pile-supported platform at the southern edge of Pier 1, replaced by the salt marsh, the footprint of each pier remains largely intact. Within the original outline of Piers 2, 3, and 5, we created cutouts in the concrete deck between fifty and ninety feet wide on their landward side, revealing open water between the pier structure and the mainland. The cutouts produced a pronounced transition from land to pier on the bridges that remained on either side of the gaps. Along with easing construction access to the added riprap revetments and expanding the tidal zone, the cutouts allow kayakers to paddle along the inner shoreline among the piers and avoid the still active shipping lanes farther out beyond the piers. They also provide parkgoers with a far more complete view of the interaction between the marine structure and the tides below. Playing basketball on Pier 2 is a singular experience because of this separation from the mainland. The courts feel more of the river than of the city, making the views to the Manhattan skyline all the more sublime.

Just north of Pier 5, the Greenway moves from solid ground onto an elevated concrete platform that runs parallel to the shore and perpendicular to the piers. A narrow, 560-foot-long cutout enabled us to preserve most of the existing deck in this location while completely transforming its relationship to the shoreline. The featureless expanse of pile-supported concrete became the Picnic Peninsula, perched between the new marina on one side and a salt marsh and constructed rock embankment on the other. Furnished with umbrellas, charcoal grills, and picnic tables made from reclaimed lumber, and with the sights and sounds of moving water on all sides, the Picnic Peninsula offers something like a seaside retreat mere steps away from the Joralemon Street entrance to the park.

On summer weekends, lone scouts show up early in the day to prepare a grill and claim a table for later arrivals of family and friends. The twenty-foot-long tables host extended family gatherings and dozens of kids for birthday parties, Fourth of July cookouts, and sunset celebrations. The design is intentionally simple, inviting parkgoers to make themselves comfortable. But even the most basic components of this venue—pavement, rails, lighting, tables, and grills—are given larger meaning by their presence on a remnant structure set within a shoreline given largely back to nature.

RIPRAP: SOFTENED ARMORING

Riprap is a low-cost, durable coastal engineering method of loosely piling angular rock into sloped embankments. MVVA has Massachusetts roots and a romantic attachment to the rock-strewn coastal areas in and around North Boston, where a combination of natural stone banks and granite quarry tailings tumble to the sea. Riprap's proven success as an industrial material, along with our instinct that the stone could be deployed artistically, made it a compelling tool for rebuilding the shoreline.

Riprap performs superbly well against moving water because the mass of the stones and the interlocking of the irregular shapes resist the forces of tides and wave action upon them. Wave energy dissipates when it contacts the porous riprap surface, and water that overtops the stone easily returns through the same gaps back into the sea. Concrete or steel seawalls often crumble during or after storms because, lacking permeability, they succumb to the hydrostatic pressure of water trapped on their inland side. For areas with wave action too powerful for a vegetated shoreline to survive, riprap is still an improvement on sheer bulkheads because it expands the tidal zone and opportunities for marine habitat. In Brooklyn Bridge Park, riprap was generously spread along the shoreline to create a middle condition between the static marine walls and the dynamic marshes and beaches.

The irregular stone serves a wide variety of functions while maintaining its same essential character, which was crucial to maintaining a material continuity throughout the park. Riprap has the versatility to serve as a simple armored bank, the cheek wall of a boat ramp, a vessel for planting soil, and informal seating. Properly positioned, it allowed for the elimination of guardrails along much of the water's edge, and it embodied the park's design principles so well that it also found its way far into the uplands. The stone edges at Brooklyn Bridge Park straddle the natural and industrial worlds, clearly placed with intention but also a part of a living environment.

DYNAMIC EDGES

Of all the shoreline transformations in the park, those that invite water into the site are most effective at heightening the experience of the river. In these few locations, the interaction of land and water is not just a line between here and there, but a destination in its own right. Twice daily, the tide changes an average of four-and-one-half feet. That shift in elevation can either be unremarkable, as when registered against a vertical seawall, or it can play out over a broad landscape, revealing the interconnected natural cycles that have evolved with it.

To decide where to bring the tides into the park, we looked at the failing bulkheads and pier decks that required replacement. We also had to consider the constraints of estuarine hydrology, potential flood events, and the location of existing infrastructure under the riverbed. All of these factors played a role in how we located, designed, and built the Pier 1 salt marsh, a dynamic and technically challenging ecosystem. This first marsh instantly became one of the most popular places in the park, proving that the urban site could support a sustained intertidal ecology, and it would inspire two other salt marshes in later phases.

A different sort of dynamic edge was inspired by Pier 4, one of the most conspicuously degraded structures on the site. Bearing no resemblance to the park's five concrete platforms, Pier 4 is a small peninsula with an 1860s wooden float bridge that once transferred rail cars to and from shore in an earlier iteration of the industrial waterfront. It had already begun to sink below the water's surface when we started working on the park, and though it could be neither salvaged nor safely occupied, we appreciated the poetry of the nineteenth-century fragment fading away as a still newer waterfront grew up around it. We separated the land at the base of the bridge from shore, intending for

The legacy of the site can be felt in the layers of past and present emerging and receding where land meets water.

it to be an inaccessible, vegetated island that could support animal habitat. The new water's edge was then converted into a sandy beach, which would also serve as the park's primary kayak and canoe launch. However, the tides forced sand into the excavated channel between so-called Bird Island and the beach, turning them back into one peninsula. The site's spontaneous reshaping has only encouraged a steady flow of visitors seeking the joy of standing at the water's edge and allowing the tide to wash over their feet. There are plans to expand the popular beach to the north and south in the years ahead.

It quickly became obvious that one beach was not enough to meet the demands of both boaters and casual visitors. To create more options for kayaking and canoeing on the river, two boat ramps were carved into the stone shoreline embankments. At the south end of Pier 1, a ramp descends parallel to the river until it meets the low tide line. In the newly opened gap between the shoreline and Pier 2, we imagined an even more dramatic entry into the water. At what is known as the Spiral Pool, a winding ramp extends well below high tide and causes the adjacent Greenway to wrap around its curving stone banks, redirecting the view of walkers and bikers toward the shore. The inlet amplifies the space between low and high tides to create a striking visual register of the water's elevation.

A CONVERSATION, NOT A COMPOSITION

To us, a park is not "composed" as if created from the top down or completed all at once. Much of our design for the shoreline was inspired by the existing conditions of the site, which varied considerably from end to end. The Greenway's relationship with the water is the clearest example of the design's responsiveness to the site as we found it, since the alignment of the meandering path reacts to the found conditions along its entire route. The Greenway lines up with the dead-straight bulkhead between Piers 5 and 6 to funnel visitors into the park. It also curves inland at Pier 2 to dodge the princess tree we found stubbornly growing on the crumbling shore, only to rejoin the water's edge in full view of the salt marsh, which is carved into that very spot because a confluence of appropriate marine and structural conditions meant it could only succeed there.

We approached the water's edge at Brooklyn Bridge Park with ambition and humility. The result is a broad range of conditions, from radical transformation to pragmatic preservation, in which each strategy amplifies the others. The remnant pile field south of Pier 1 strikes a more impressive figure set against the salt marsh's bronze cordgrass and rocky embankments. The dynamic curvature of the Spiral Pool is more arresting for its adjacency to the rectilinear edge of Pier 2 across the way.

At the same time, all of these edge conditions are grounded in practicality. They are responses to the site and its potential. What deteriorated and what persisted? What was once living here and what could we bring back? What would people feel when they had the chance to touch water again? The history of the site is now legible in almost every aspect of the park's design and construction. It is not a legacy that has to be explained, but one that can be felt in the layers of past and present emerging and receding where land meets water. Some pieces willfully hang on from a past era, determined to live again; some clearly signal a new future; and some still shift at your feet.

ABOVE *Demolition of the concrete deck on the south side of Pier 1 revealed the field of piles below, 2009.*

FOLLOWING PAGES *An aerial view of Pier 1 as built, including the pile field, salt marsh, and Harbor View Lawn, 2021.*

A COMPLEX EDGE

The water's edge was invigorated with a mix of preserved industrial structures, naturalized edges, and stabilized embankments that protect the park from flooding and wave action.

LEFT *The footprint of the site before and after structural interventions on the shoreline.*

RIGHT *View of the platform structures, riprap banks, and beach along the water's edge between Pier 2 and the marina.*

A footpath on the riprap edge of the Pier 3 cutout.

PRESERVED AND REPURPOSED

Placing landscapes and recreational programs within the historic footprints of the cargo sheds left the piers' broad perimeter walkways intact. Cutting away portions of the decks on the inland side of Pier 2 and 3 created passageways for kayaks.

RIGHT *The preserved perimeter walkway on Pier 3 allows for active and passive activity for visitors.*

FOLLOWING PAGES *The Pier 3 perimeter walkway.*

PIER 5

89

SOFTENED ARMORING

Failing bulkheads were replaced with more resilient stone revetments that allow flood-waters to seep back into the East River.

LEFT *The beach at Pier 4 looking toward the Pier 3 uplands.*

RIGHT *Pier 4 beach with the Pier 5 Picnic Peninsula beyond.*

PIER 3

PIER 3

DYNAMIC EDGES

Restored salt marshes and new beaches encourage parkgoers to stand in the East River and witness its tidal change and biodiversity.

PIER 2

FLOOD LEVEL,
HURRICANE SANDY, 2012

HURRICANE SANDY

In October 2012, Hurricane Sandy caused a 13.88-foot tidal surge that inundated the half-finished park with saltwater. The landscape suffered minimal damage.

ABOVE *The indundated landscape at Main Street Park in DUMBO.*

RIGHT *The salt marsh at Pier 1.*

LEFT *The amphitheatre and beach at Main Street Park.*

FOLLOWING PAGES *Access to the water's edge under the Manhattan Bridge, between Main Street and John Street*

The salt marsh at John Street Park cradled in protective riprap and salvaged building footings discovered during construction.

THE SALT MARSHES

Three salt marshes have been carved into the park's shoreline. The smallest is cradled in the stone revetment behind the Picnic Peninsula at Pier 5. The last to be built is a variation of a meandering intertidal zone and is situated just north of the Manhattan Bridge. But the first and largest marsh, just south of Pier 1, best shows the unique beauty and ecological value that even a pocket-sized salt marsh can yield in a designed landscape.

The East River is not, in fact, a river. It is a tidal strait that flows from the push and pull of its dual connections to the Atlantic Ocean: the Long Island Sound to the north and the Upper New York Harbor to the south. These conflicting currents can create treacherous boating conditions as the direction of flow changes with the tides. Before the growth of the city, erratic currents and occasional storm surges had long been moderated by a vast network of now-lost beaches, dunes, mudflats, and salt marshes.

Salt marshes naturally occur in narrow bands of protected saline shoreline that are inundated twice daily by the tides. On open beaches, tidal cycles can be forceful and erosive. But salt marshes dissipate wave energy, allowing decomposing organic matter to stay in place and thus feed a complex food web. Salt marshes are near monocultures of smooth cordgrass, *Spartina alterniflora*, which spreads via underground rhizomes and forms a dense mat. Microorganisms break down the decaying vegetation and support food webs of diverse coastal wildlife. In the New York area, this includes the remarkable variety of harbor herons that nest and forage in salt marshes, activities that also nourish the cordgrass.

Marshland was loathed in the early industrial era, seen as an obstacle to the creation of valuable waterfront property and the functioning harbor of a developing city. The silty shallows and low-slung uplands of the intertidal zone conflicted with the deep channels and vertical edges at the land-water interface that were required to transfer cargo. Filling the soft contours of New York City's riverine edges eradicated roughly 20,000 acres of salt marshes in just a few hundred years. This enduring injury to New York City's natural landscape is hard to appreciate today. We were determined to reestablish a measure of the site's original biodiversity and environmental resilience in a way that would be restorative for people, too.

To recreate a successful intertidal habitat in postindustrial conditions required an inventive attitude, letting go of any preconceptions about what urban and coastal ecologies should be. The needs of the new salt marshes were woven into found fragments of the industrial past. Taking some inspiration from Matt Urbanski's childhood experiences on the New Jersey shore, we went on field visits to nearby restored and preserved coastlands. Across the harbor at New Jersey's Bayonne Golf Club, we studied a newly constructed salt marsh. In 2005, Eric Rothstein, a hydrologist and restoration ecologist, led site visits to Pelham Bay Park in the Bronx, where a 200-acre network of natural marshland drove home the beauty of tidal wetlands.

When looking at constructed and functioning marshes, the key takeaway is the significance of a marsh's elevation, which permits the tidal rhythm of inundation. Naturally occurring salt marshes will develop over hundreds of years and settle into a dynamic relationship with the water, but a constructed marsh needs to be precisely calibrated to the tides on day one to survive. If the elevation is too low, most grasses become immersed in saltwater beyond their tolerance. If the elevation is too high, the tides can't deliver the sediments that the marsh requires to support growth. We had gathered vital strategies for shoreline restoration through field research, but applying the findings to the site presented unique challenges.

The East River's naturally erratic currents are made more complicated by existing infrastructure, wake from boats, and a changing climate. Our first proposed location for a salt marsh was in the large cove in front of the Empire Stores, but that site did not provide adequate shelter from the particularly strong currents in that section of the river. Besides, the New York State Department of Environmental Conservation (NYSDEC) warned that gaining their approval to add fill to the waterway was unlikely.

Soon after, we found that the south side of Pier 1, which was being considered for a small boat marina, was prone to sediment buildup as the river currents deflected off the pier. The same phenomenon that made the site unsuitable for a marina made it ideal for a salt marsh. The space between Piers 1 and 2 formed a protected cove, and the ongoing sediment deposition would naturally nourish a new marsh. In addition, the Stoplight Plan indicated that

We were determined to reestablish biodiversity and environmental resilience in a way that would be restorative for people, too.

existing marine walls were failing in this location. Although sediment was already collecting at the south end of Pier 1, building a marsh out into the open water would have again raised objections from NYSDEC. They would, however, allow us to remove the decking and carve away at the shoreline, expanding the extent of open water.

Although nine acres of Pier 1 consisted of a concrete deck supported by solid fill, the three-acre portion of its southern end sat on top of deteriorating timber piles much like the other piers. After the concrete deck was removed, the grid of piles that remained comprised a ready-made system for attenuating wave action and blocking ice floes that would damage the marsh.

To form a new water line at the land's edge, we carved a gentle upward slope from a quarter acre of fill below the pier. We kept a row of massive concrete columns that mark the original transition from platform above the water to landfill, and we added a narrow stone revetment behind it to further insulate the marsh from wave action that could wear away the soil. The riprap shoreline on the inland side blends the marsh seamlessly into its surroundings. The convergence of edge conditions is striking in its complexity, with rugged layers of stone, plants, and water comprising the foreground to the New York harbor beyond.

The earth fill that was uncovered below grade was far from the well-draining medium needed for a constructed marsh, so we turned to manufactured soils to support the ecology of the intertidal zone while ensuring its long-term performance. Our research into salt marsh restoration projects suggested that a sandy soil would be the most supportive for our site rather than the silty soil typical of natural marshes. To allow the marsh to drain completely at low tide, a coarser growing medium with minimal water retention ability is more important than an absorptive, nutrient-rich soil, as tidal waters are already replete with nitrogen and phosphorus. The sandy medium was installed over a porous filter stone as an added measure to ensure excellent drainage when the water recedes.

In this new growing material, we planted thousands of tiny, smooth cordgrass plugs. The timing was tricky. Cordgrass is highly adaptable, but the initial establishment of a stand of plants must be coordinated carefully. The small plugs were custom grown at the Pinelands Nursery in New Jersey, where they were acclimated to the same salinity levels found on the shores of Brooklyn Bridge Park. Because *Spartina alterniflora* is a warm-season grass, the plants could be installed neither too early in the year nor too late, which would open the possibility of cold weather returning before they could root in. After surviving its first season, the cordgrass showed persistence against hungry geese and the extreme flooding during Hurricane Sandy in 2012, which flattened the grasses but left their root systems unscathed. Although the marsh needs occasional raking to remove floating debris and mats of dead grasses, it otherwise requires little maintenance.

The Pier 1 salt marsh covers just 8,000 square feet. It is tiny compared to natural marshes but functions the same way, harnessing the tidal exchange between water and land to help sustain a rich matrix of plant, animal, and nutrient interactions. It gives people a chance to experience one of the many habitat types that together contribute to the park's biodiversity. Several estuarine species, including bivalves and crustaceans, have taken up residence in the marsh soil and among the grasses, and the site has become an important stopover for migratory species like shorebirds moving between larger habitats in Jamaica Bay and the New Jersey Pine Barrens. The marshes are favorite destinations for educators who bring classes to the park, as their small size, accessibility, and simple composition make them highly approachable living classrooms. Having easily recovered from Hurricane Sandy, the park's restored tidal zone offers a reason for optimism about the use of natural systems to protect coastal cities from flooding.

The view across the harbor is a mesmerizing sight from a seat on a rough granite block on the edge of the Pier 1 salt marsh, looking out over the smooth cordgrass bending in the wind and catching rays of sunlight in a barely perceptible fringe of seedheads. Even with the spectacle of Lower Manhattan in full view, the eye is drawn to the soft patch of green where tidewater passes through the stems. To New Yorkers who justifiably worry about sea level rise, floods, and biodiversity loss, the park's salt marshes are powerful reminders that the East River is a living waterway with much to offer the city.

LEFT *The concrete deck of Pier 1 severed from the solid ground that would hold the salt marsh, 2009.*

RIGHT *The remaining pile field and completed Pier 1 salt marsh, 2010.*

ESCAPE IN THE CITY

PIER 6

PIER 6

PIER 5

COME SAIL WITH US
IX 514

PIER 5

PIER 5

PIER 3
PIER 2
MetLife
NATIONAL

PIER 4

14

PIER 3

PIER 3

THE GREENWAY

PIER 2 UPLANDS

PIER 2

PIER 2

PIER 2

PIER 1

PIER 1

PIER 1

PIER 1

PIER 1

PIER 1

PIER 1

PIER 1

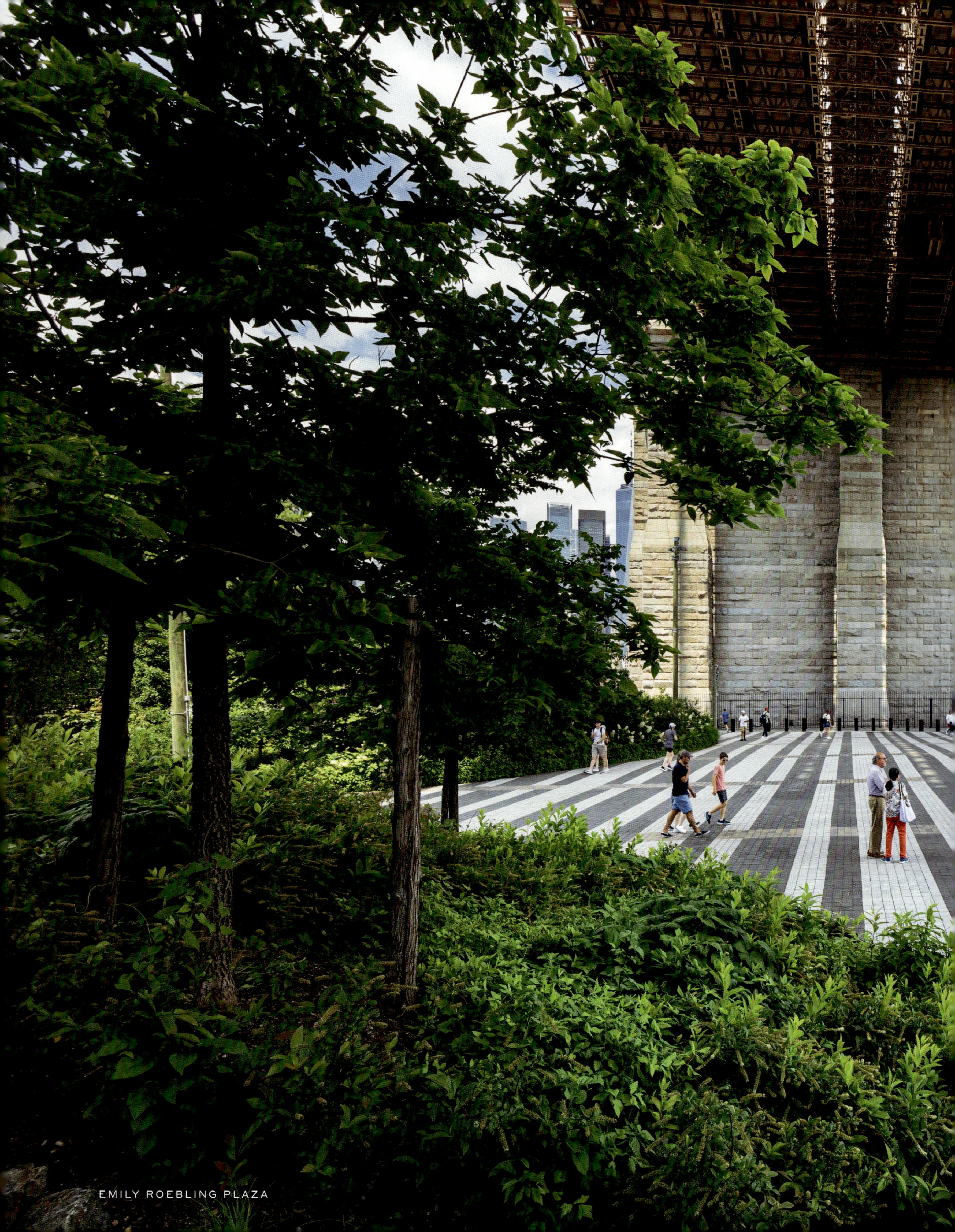

EMILY ROEBLING PLAZA

EMPIRE FULTON FERRY

MAIN STREET

AFTERWORD

AMANDA HESSER

I first heard of Brooklyn Bridge Park years before it existed. In 2002, when my then-fiancé, Tad, and I trudged around Brooklyn Heights, searching for an apartment where we could imagine spending our future life together, our broker mentioned that the piers on the waterfront were going to be transformed into a sumptuous park.

Until then, we'd found her trustworthy. But having seen the squalid piers from the Promenade—idle trucks, decrepit pilings, rusting containers, piles of refuse that had clearly been neglected for decades—her promise seemed like a total sales job. This urban wreckage was beyond repair. And so it remained after we'd found a place in the Heights, for many years.

The city was still recovering from 9/11, and repair and stability were higher priorities than urban renewal. Only occasionally were the bleak gray piers animated by public projects such as Olafur Eliasson's *Waterfalls* and a floating pool parked near Pier 5, complete with heaps of sand to simulate a beach. The rest of the time it was just the same old eyesore that was part of a New Yorker's birthright: the part you'd edit out of any photo of the Statue of Liberty, or the Manhattan skyline.

Then, in 2009, we saw the landscape begin to change. Bulldozers attacked Pier 6, and the bleakness suddenly vanished, exchanged for a set of densely planted pockets inscribed with pathways. Trees and bushes popped up, swarming a maze of sprinklers, swings, and climbing structures that our children would soon roam through and scale. There were beach volleyball courts and even a pizza kiosk. From just one pier on the East River emerged the promise of a truly great public park.

Our real estate agent gained our sudden retrospective esteem. And assessing the park's daily progress became our favorite new sport. Pier 1 reopened in 2010 with a large grass lawn where we would picnic, a playground where our toddler twins would frolic, and a network of walkways through which I'd fast walk, delighting in the lush green space. Now that the bookends of the park had been established, what would go in the center?

The great pleasure of watching the park bloom was its incremental evolution. So many big urban renewal projects are completed in one fell swoop. Brooklyn Bridge Park developed pier by pier, swing set by swing set, field by field. Because no central news source—at least, not one we were aware of—was reporting on the plans and their timeline, every new tree or mound that appeared felt like a surprise. We remained blissfully unaware of the community board battles and funding delays and simply watched the rejuvenation unfold. My son and I regularly peered over the Promenade railing to scour the earthworks below for any developments, then took walks through the makeshift passageway from Piers 1 to 6 to check out each of the changes at eye level.

Most large construction projects keep gawkers like us out. But this project—perhaps because it was a collaboration of public and private constituencies—invited us in to explore, and in turn, gained our devotion. It used any scrap of free space to draw the public into the park, even in its most unkempt state. A raw gravel parking lot between Pier 5 and Furman Street hosted Smorgasburg and Photoville, an annual exhibit celebrating emerging photographers from around the world.

Public art became a fixture of the park. An interactive water sculpture called *Appearing Rooms* by Jeppe Hein was installed on Pier 1 in 2015, bringing kids much joy and soggy clothing. In 2016, amid a contentious presidential election, Martin Creed's massive rotating sculpture of the word "Understanding" dominated Pier 6. I often walked down to see it, to remind myself of the word's many meanings.

Along the makeshift road winding through the park, they installed a Pop-Up Pool, which became a magnet for local kids, including our own. When our son became obsessed with soccer, Pier 5 reopened as if on cue with three full soccer fields in the center and wash basins at the pier's end for the local fishermen—a detail that underscored how the park's designer, Michael Van Valkenburgh, worked to include everyone. I would often remind my son that playing soccer on fields that appear to be floating on water, with the Statue of Liberty and the Freedom Tower as a backdrop, is not most kids' childhood experience. This fell on deaf ears, naturally; he was just happy to have fields close to home, and often went down to shoot penalty kicks on the empty goals, dreaming of glory.

The soccer fields would later serve as our outdoor oasis during the 2020 lockdown. After long days inside our apartment, we'd venture out to Pier 5 to do calisthenics, kick a ball, and fast walk. Staying together in the open air of the pier felt safe; we were as grateful for this as we were to see other people.

Soon after Pier 5 opened, Pier 2 did, too, drawing basketball players by the hundreds. The pier somehow managed to squeeze in bocce courts, which now also double as pickleball courts, as well as a roller rink and a small turf field for practicing ball control and one-on-one games. My favorite pier, Pier 3, opened last. It's the most straightforward of the piers, a gently rolling stretch of lawn and trees, surrounded by water.

It is an understatement to say that Brooklyn Bridge Park has changed Brooklyn. It's become North Brooklyn's center of gravity, energizing the local neighborhoods by drawing visitors from around the city, and around the world, and providing a much-needed space for relaxation and play. Because you can practically only enter the park from either end, you're led to experience more of it as you pass through it to reach your destination. The park's spatial limitations—its narrowness, a slender green capillary connecting DUMBO to Red Hook—are actually a strength. The park serves as a passport for the tri-border area, connecting the Heights to its neighbors in a seamless web.

My own family grew up with the park. Our twins first enjoyed the park as toddlers when it, too, began standing on its own. They now meet their friends there for pickup games and picnics. For my husband and me, the park has become an essential recharging station and a safe space for intimate conversation. Nearly every day, we walk our dog down the Squibb Bridge to Pier 1 and then through the park to Pier 6 and back up Joralemon Street to the Heights and our home. We talk about our days, our kids, our worries. Sometimes we might not even talk at all. We just take in the surrounding greenery and the majestic view that seems, still, like a miracle.

TEAM & COLLABORATORS

AGENCIES

CITY

New York City Mayor's Office
Department of Parks and Recreation
Department of Transportation
Economic Development Corporation
Department of City Planning
Metropolitan Transportation Authority
NYC Council Members (various)
Department of Buildings
Department of Environmental Protection
Brooklyn Borough President's Office
Emergency Management Office
NYPD
FDNY

STATE

Empire State Development Corporation
State Historic Preservation Office
Office of Parks, Recreation and Historic Preservation
Department of Environmental Conservation
NYS Senators and Representatives (various)
Public Authorities Control Board
NYSERDA

FEDERAL

United States Army Corps of Engineers
United States Coast Guard

COMMUNITY

Brooklyn Bridge Park Conservancy
Brooklyn Bridge Park Community Advisory Council
Brooklyn Heights Association
Community Board 2
Cobble Hill Association
Community Board 6
Brooklyn Greenway Initiative

CONSULTANTS

Accu-Cost Construction Consultants, Inc., Cost Estimating
AECOM (formerly DMJM + Harris), Civil, Marine and MEP Engineering
Altieri Sebor Wieber, LLC, MEP Engineering
AltPower/RELAB, Renewable Energy Consultants*
Architecture Research Office (ARO), Architecture
Battle McCarthy, Environmental Engineers*
Cerami & Associates, Acoustical Engineering
Clark Wolf, Food and Restaurant Consultants*
Constructive Strategies, Government and Community Relations*
Cooper Robertson & Partners, Urban Design*
Domingo Gonzalez Associates, Inc., Lighting Design
Ducibella Venter & Santore, Risk and Protection*
E2 Project Management, Structural Engineering
Easton Architects, Architecture
Eng-Wong Taub & Associates, Transportation Planning
F2 Environmental Design, Soil Scientist
Fluidity, Fountain Consultant
Gensler, Architecture and Conditions Surveying*
Great Eastern Ecology, Ecology
Greenberg Consultants, Urban Design*
Henry Bardsley, RFR, Infrastructure Engineers*
HR&A Advisors, Economic Consultants
James Carpenter Design Associates, Architecture*
James Macdonald Consulting, Code Review
Kenny & Kahn, Code Consultants
Lodging Investments Advisors, LLC, Hotel Consultants*
Margie Ruddick Landscape, Ecological Landscape Planning*
Maryann Thompson Architects, Architecture
Mathews Nielsen Landscape Architects, Landscape Maintenance Cost
Nitsch Engineering, Stormwater Reuse
Northern Designs, Irrigation Design
OPEN, Graphic Design
Paulus, Sokolowski and Sartor, Architecture
Pentagram Design, Graphic Design*
Pine and Swallow Associates, Soil Scientists
Richmond So Engineers, Inc., Structural Engineering
R.J. Van Seters Company, Fountain Consultant
Sam Schwartz Engineering, Transportation Planning
Sherwood Design Engineers, Civil Engineering
Steven Handel, Ecology*
Susan Fine, Economic Consultant*
Urban Strategies, Urban Design*
Ysrael A. Seinuk, PC, Structural Engineering
WXY Studio, Park Maintenance Planning

*Master Plan phase only

MVVA TEAM

Ahn, Michael
Aihara, Megumi
Alter, Curtis
An, Clayton
Andersson, David
Andrews, Jonathan
Angell, Marissa
Arato, Peter
Archer, Hillary
Armesy, David
Askin, Samuel
Bachman, William B.
Ballhaus, Johanna
Barner, Timothy
Bazzle, Bradley
Benintende, Emma
Bennett, James
Benson, Jason
Benton, Charles
Berrizbeitia, Anita
Bird, Matthew
Bivona, Joseph
Blondell, Evan
Bolstad, Jennifer
Bonifaci, Emily A.
Boyd, Andrew
Brazelton, Pauline
Brightman, Eric
Brown, Michael
Budzinski, Neil
Burcheri, Aldo
Burns, Marni
Button, Drew
Camuti, Liz
Canter, Alexis M.
Carboneau, Jean-Paul
Cassidy, William
Cassini, Alexander
Cedar, Michael J.
Chase, Austin
Chastka, Astrid R.
Chiari, Ana
Chin, Ting
Chiong, Maximillian
Cho, Sangyun
Choi, Danielle
Clark, Sarah V.
Clifford, Martha
Coogan, Matthew
Cook, Astrid
Corriel, Sean
Counts, Christopher
Cregg, Emily
Cronin, Aidan
Culbreth, Zachary
D'Aurizio, Max
Daniels, Inge
Dannin, Isadora
David, Simon
Davis, Alexandra
Dawley, Cole
Debnath, Shibani
Deng, Jing
Diamond, Sarah
Dimitri, Alexandra
Ding, Yu
Divekar, Parth
Doone, Ryan
Dropkin, Ethan
Egnatz, Joshua
Elachi, Lauren
Elkovitch, Nicholas
Ellickson, Stacie
Fabbri, Sophie
Farmand, Sahar
Feeney, Corryn
Foldes, Kenzie
Friend, Alexia
Fritz, Geoffrey
Fryatt, Timothy
Gao, Yufan
Garcia Simon, Paloma
Gates, Christopher
Gazzo, Timothy
Georg, Hana
Giarratana, Jennifer
Gidding, Can
Gifford, Kenneth
Gilbert, Elizabeth
Girard, Matthew T.
Glazer, Jacob
Gleeson, Rachel
Godner-Abravanel, Alana
Goksel, Ayse
Goodrich, Catherine
Goodrich, Scott
Gordon, Emily
Gregg, Jesse A.
Gulick, Nicholas
Gutterman, Andrew
Guzzetta, Max
Haddad, Rana
Haffner, Edward C.
Haines, Ernest
Harris, Luke
Hartsfield, John
Harvey, Kerrie
Hassell, Alyssa
Heflich, Adrienne
Heilner, L. Justine
Herbert, Richard
Hill, Aaron
Hindle, Richard
Hirsch, Brian
Hirschbeck, Mark
Hobdy, Alex
Holland, Elspeth
Holland, Laurence
Hong, Zaneta
Hopper, Albert
Hoyt, David
Isaza-Figueroa, Intiya
James, John-Michael
Javellana, Austin
Jirasiritham, Kate
Johansen, Anna
Johnson, Monique
Johnston, Haley
Jones, Dylan
Jostol, Edyth
Kahler, Leah
Kantner, Ananda
Kelly, Erin
Kelly, Scott
Kim, Hyung Nyun
Kim, John L.
Kim, Phillip
Kirby, Tim
Kochuba, Christian
Koren, Stephen J.
Krob, Tyler
Kumar, Ishaan
Ladner, Conners
LaRocca, Jason
Larsen, Kate
Lauder, Reagan
Ledley, Brendan
Lee, Wilson
LeJeune, Blair
Levy, Izabela
Li, Qiyi
Li, Ye
Liang, Annie
Liao, Tzufen
Liebmann, Amy
Linck, Preston
Liparulo, Matthew
Liss, Katherine
Little, Megan
Liu, Pam
Loomis, Stefanie
Lovinger, Leor M.
Lucius, Kristopher
Luckmann, Alexander
Luegering, Michael
Luhrmann, Fiona
Lynn, Lauren
MacLeod, Michaela
Madson, Ryan S.
Malik, Karol
Marcatili, Sandra
Martin, Jeffrey
Maurer, Chris
McCullough, Angus
McMahon, Matthew
Micsak, Robert
Miller, Leo
Mishima, Yoshiki
Mohr, Tyler
Moin, Sahar
Ness, Lukas
Newsome, Jason
Nicholson, Jesse
Noone, Stephen
O'Brien, Killian
O'Connor, Kyle
O'Reilly, Clare K.
Ohly, John
Ormston-Holloway, Michael
Parrilli, Nicholas
Pattin, Mark A.
Pelletier, Grace
Pevzner, Nicholas
Pham, Linh
Phillips, Esther Parker
Phillips, Paul
Pindyck, Jennifer
Pirokka, Michalis
Prado, Jason
Primerano, Kaylynn
Prince, Erik D.
Proft, Joanne
Pruitt, Charles
Rafuna, Armend
Raganit, Estello
Ravestein, Adam
Reimer, Christina L.
Reynolds, Sally
Rigney, Tamara M.
Rock, Robert
Rodriguez, Charlotte
Rose, Michael
Ross, Natalie
Roura, Martin
Saltarella, Michael
Satterlee, Jane
Schneider, Max H.
Schuette, Madeline
Seck, A. Paul
Seelman, Brett
Shepard, Gullivar
Shively, John
Siebenmorgen, Jason
Silver, Elizabeth
Slob, Martijn T.
Smith, Charles
Smith, Michael T.
Solano, Laura
Song, Humbi
Spangler, Alec
Stadecker, Gabriel
Strange, Matthew
Streeb, Scott
Strobel, Adam
Sweeney IV, Herbert T.
Tacey, Carlin
Tang, Dorothy
Tao, Diana
Tietje, William
Tominaga, Ryoma
Toomey, Kasey R.
Traut, Taylor
Trevethan, Nathan
Tsai, Faustina
Urbanski, Matthew
Vajda, Nicole
Van Valkenburgh, Michael
Vargas, Chelsea
Voelkel, Michael
Volkmer, Kayla
Wachirabuntoon, Parawee
Wai, Antonia
Walsh, Cara
Wang, Jing
Welch, Sarah M.
Wellington, Benjamin
Wiggins, Martin
Wilson, Michael
Wisniewski, Andrew
Wojtas, Kevin
Wu, Angela
Wyatt, Katherine
Wythoff, Erin
Zack, Jonathan

THE BBP OPERATIONS TEAM

The Brooklyn Bridge Park Corporation Team brings together a wide variety of horticulture, maintenance & operations, and administrative staff who are devoted to rigorous and sustainable stewardship. Here is a snapshot of this dedicated team in March 2023.

TOP ROW, FROM LEFT TO RIGHT Dwens Pierrot, Maren Cadwallender, Paul Sutter, Devon King, Corey Rivers, David Lowin, Reza Beniamer, Evelyn Manlove, Kim Cass, Christina Severin, Dylan Barrett Smith, Stephanie Pace, Kelli Wilcox, Ben Theriault, Curt Wilson, Eric Landau, Jonah Goldstein, Kyle Beyer, Dana Glass, Sean Reynolds, Rashid Poulson, Jeffrey Sandgrund, Alanna Rios, Joel Bermejo, Minnow Johnson

BOTTOM ROW, FROM LEFT TO RIGHT Conner Bush, Bella Ciabattoni, Ariel Ries, Jorge Ferreira, Pawel Pieluszynski, Jojo Fernandez, Eliza Perkins, Sarah Krauss, Amy Kleitman Lojek, Onika Shelby, Michael Cartegena, Richard High, Richard Shepherd

ACKNOWLEDGMENTS

This book complements two earlier accounts, both by true insiders, of the battle to turn the abandoned East River port into a park. In *A History of Brooklyn Bridge Park: How a Community Transformed and Reclaimed New York City's Waterfront* (2016), Nancy Webster, Executive Director of the park's Conservancy, and journalist David Shirley chronicle efforts starting in the 1980s to claim the site for the public. Joanne Witty, the first President of the Local Development Corporation, and author Henrik Krogius collaborated on *Brooklyn Bridge Park: A Dying Waterfront Transformed* (2016), which focuses on the numerous obstacles the community overcame to bring the park to life. MVVA's work has built on the story of activism retold in these two volumes. It is fitting that our book should follow theirs.

We are grateful to Alec Spangler, a professor of landscape architecture and MVVA alumnus, for contributing so much of the writing in this book, and to Fred Bernstein, Joe Bargmann, and Neil Budzinski for their invaluable efforts to advance the text over the past two years. Elijah Chilton managed the book with great dedication for well over a year. He and Michael Van Valkenburgh were the driving forces behind the development of the essays.

We are indebted to Elizabeth White, our editor at The Monacelli Press, for her unwavering support for this project. Special thanks to Julie Bargmann, whose approach to our shared discipline makes her the perfect voice to kick off this book, and to Amanda Hesser for closing the volume with her memories of the park's progress as a resident of Brooklyn Heights for two decades.

The design of the book was conceived and executed by Cara Walsh. A landscape architect and graphic designer, she tells a story across the pages in a way no one else could. Thank you also to Chris Donohue, Nate Trevethan, and Kate Jirasiritham for supporting Cara's excellent work and guiding our efforts to represent the park.

Images are as important as words in telling the story of Brooklyn Bridge Park. We are fortunate to be able to show the park through the eyes of profoundly talented photographers. Thank you to Elizabeth Felicella, Etienne Frossard, and Alex MacLean, who have captured the park in progress for over twenty years, and to Julienne Schaer, Scott Shigley, Lexi Van Valkenburgh, Izabela Riano-Levy, and Conor Coghlan. So many colleagues, clients, employees, and friends have contributed to the shape and feel of the book, particularly its essays. Thank you to John Alschuler, Peter Arato, Julie Bargmann, Amelia Boyd, Ethan Carr, Danielle Choi, Nilda Cosco, Aidan Cronin, Nik Elkovitch, Olivia Gee, Liza Gilbert, Eva Gildea, Matt Girard, Rachel Gleeson, Ken Greenberg, Sophie Kaplan-Bucciarelli, Alexander Luckmann, Chris Matthews, Brad McKee, Rebecca McMackin, Robin Moore, Heidi Natura, Steve Noone, Catherine Seavitt Nordenson, Oliver Oglesby, Fletch Phillips, Pawel Pieluszynski, Elizabeth Barlow Rogers, Eric Rothstein, Matthew Schaeffler, Jason Siebenmorgen, Suzanne Turner, and Andy Wisniewski.

It is an even greater challenge to recognize all the people who contributed to the planning, design, construction, and care of Brooklyn Bridge Park. It takes an unimaginable number of deeply devoted engineers, artisans, growers, masons, ecologists, woodworkers, metalworkers, gardeners, financial experts, and more to create and sustain a park like this one. Special thanks to Hector Ducci, who guided the hard work of salvaging and repurposing the timbers that became the park's benches. Thank you to John Kelly and his brilliant crew at Kelco Construction, who built most of the park over almost fifteen years.

The maintenance and repair efforts of the Brooklyn Bridge Park Corporation have been enormously rewarding and instructive. We are lucky to benefit from gardeners and others who seek our advice while making their own mark on the landscape every day, directly contributing to the success of the park. The early years of plant establishment were guided by the skillful Rebecca McMackin as the first Director of Horticulture. Thank you to Rashid Poulson, Rebecca's successor since 2022, and to horticulturalist Evelyn Manlove, who walk the park with Jason and me four times a year.

Our team is grateful for the park's widespread public support and critical reception since the completion of the master plan. Thank you to Anita Berrizbeitia, who curated one of the earliest exhibitions of our work on the park, *Between Form and Circumstance*, at the Harvard Graduate School of Design in 2006. Thank you to Rosalie Genevro and the Architectural League of New York, who sponsored a public presentation on the park at the League in 2011.

Finally, thank you to Cara Walsh and the Van Alen Institute, whose 2012 exhibition, *Immensity + Intimacy*, helped convey Brooklyn Bridge Park's important role in the future of urban waterfronts.

Thank you to the many staff who help keep the park clean year-round, many of whom are hired through the Doe Fund, a transitional work-training program that connects previously incarcerated and homeless New Yorkers with maintenance positions.

Our team has benefited from the continuous support of four successive New York City Mayors, including Rudolph Giuliani, Michael Bloomberg, Bill de Blasio, and Eric Adams. In the critical early phases of planning, we were enormously lucky that Mayor Bloomberg and Deputy Mayor Patricia Harris arranged to transfer ownership of the park from the State of New York to the City of New York, and created the Brooklyn Bridge Park Corporation. Each of the BBPC Presidents—Jim Moogan, Wendy Leventer, Regina Myer, and Eric Landau—have been pivotal in shaping the park's identity and empowering MVVA to do the same.

Thank you to Jeff Sandgrund, Chief Operating Officer at the BBPC, who has been a part of the park since the first path was paved, and to David Lowin, the former Executive Vice President. And for their support in planning and design, we'd like to thank Chip Place, Jennifer Klein, Leigh Trucks, Pat Kirshner, and Lindsey Ross.

We are indebted to the advocates who toiled and schemed to make this park happen, including Fred Bland, Otis Pearsall, Scott Hand, John Watts, David Offensend, Joanne Witty, Hank Gutman, Mary Ellen "Mickey" Murphy, and many others. Their belief in the project never abated throughout the many years of its realization, and their advocacy continues to this day. Thank you to all past and present members of the BBPC Board of Directors who have helped keep their vision alive.

When the Corporation was still in its infancy, we depended on the guidance of New York City experts with deep knowledge of local parks and governance. Thank you to NYC Department of Parks & Recreation Commissioner Adrian Benepe and to the staff at the Battery Park City Authority, particularly Executive Director Tessa Huxley and Director of Horticulture Eric "T" Fleischer, for attending countless lunchtime meetings at our office and providing us with the direction we needed.

Brooklyn Bridge Park was built almost entirely with public funds from the City and State of New York. Thank you to the taxpayers, whose contributions have made a seemingly impossible project a reality and brightened the lives of millions of visitors each year.

First published in the United States by
The Monacelli Press, a division of Phaidon Press, Inc.

Library of Congress Control Number: 2023940672
ISBN: 978-158093-617-0

Design: Cara Walsh
Coordination: Elijah Chilton, MVVA

Printed in China

The Monacelli Press
A Phaidon Company
111 Broadway
New York, New York 10006

PHOTOGRAPHY CREDITS

Conor Coghlan: 104

Elizabeth Felicella: 11, 16-17, 26-27, 30-33, 40, 42, 45, 67, 69, 73, 90-92, 94-95, 116-117, 128, 129, 150, 153, 157-161, 163-165, 173, 184-187, 210-211, 222-223, 227, 232-235, 240, 244-245, 248-249, 251, 262-263, 265, 268-269, 282-283

Etienne Frossard: 4-5, 8, 24-25, 41, 46-47, 64-65, 66, 70, 71, 86, 98-101, 122, 124-127, 130-133, 140-141, 154-155, 162, 168, 169, 172, 174, 201-209, 212, 213, 236-243, 246-247, 250-259, 264-267, 270-281, 284-291, 294-295

Alexa Hoyer: 43

Rafil Kroll-Zaidi: 216

Alex MacLean: 14-15, 36-39, 48-49, 62-63, 74-77, 85, 87, 96-97, 110-111, 119, 120-121, 135, 139, 142-145, 170-171, 188-189, 193, 194-195, 197, 200, 218-219, 230-231, 302-303

Izabela Riano-Levy: 228-229

Julienne Schaer: 2-3, 82-83, 123, 214-215

Scott Shigley: 6-7, 59, 198-199, 240, 241, 260-261, 264, 265

Scott Stowell: 28

David Sundberg/Esto: 105

Lexi Van Valkenburgh: 20-21, 93, 106, 107, 108-109, 175, 176-177, 180-181, 216-217, 220-221, 251, 298-299

IMAGE CREDITS

Page 152: Eric Drooker, Last Stop for *The New Yorker*, August 5, 2002